Spectrochemical Analysis by Atomic Absorption and Emission
2nd Edition

Spectrochemical Analysis by Atomic Absorption and Emission

2nd Edition

L. H. J. Lajunen and P. Perämäki
University of Oulu, Finland

RS•C

advancing the chemical sciences

ISBN 0-85404-624-0

A catalogue record for this book is available from the British Library

Published by The Royal Society of Chemistry,
Thomas Graham House, Science Park, Milton Road,
Cambridge CB4 0WF, UK
Registered Charity Number 207890

For further information see our web site at www.rsc.org

Typeset by the Charlesworth Group, Wakefield, UK
Printed by Athenaeum Press Ltd, Gateshead, Tyne and Wear, UK

Preface

Atomic spectrometric techniques and ICP-MS are frequently used in trace element analysis in many laboratories. Hence the knowledge about these instrumental methods is essential for those who use these techniques as well as for those who utilize the results obtained.

The present book describes the basic theory of atomic spectroscopy and deals with the most common techniques (flame AAS, graphite furnace AAS, Plasma AES, AFS and ICP-MS) that are used for trace element analysis in academic institutions, research laboratories, environmental laboratories, etc.

Sample introduction is a critical step that affects the sensitivity and possible interference effects encountered in various methods. In principle, similar sample introduction methods can be used with all techniques, and therefore most common sample introduction techniques, as well as instrument components, are reviewed.

Most common possible interference effects involved in different analytical techniques are presented. Sample preparation methods are also briefly overviewed in order to underline that sample preparation must be chosen according to the analytical information needed.

In addition, most common hyphenated analytical techniques are also included since speciation analysis with highly selective atomic spectrometric detectors and ICP-MS is becoming more and more important for instance in environmental research.

The present book can be used to teach these analytical methods both for undergraduate and graduate students. The book suites also well for those who already use these techniques and want to learn more about these methods and theories.

Since the publication of the first edition "Spectrochemical Analysis by Atomic Absorption and Emission" in 1991, various techniques in analytical atomic spectroscopy have undergone some development. The

v

present book is an updated and revised edition of "Spectrochemical Analysis by Atomic Absorption and Emission" by L.H.J. Lajunen.

Lauri H.J. Lajunen and Paavo Perämäki
Oulu, April 2004

Contents

Introduction

1.1 HISTORICAL

The first spectroscopic observation was made by Newton in 1740. He discovered that the radiation of white light splits into different colours when passing through a prism. In the middle of the nineteenth century metal salts were identified by means of their colour in the flame. The first diffraction grating was introduced by Rittenhouse in 1786.

In 1802 Wollaston discovered, in the continuum emission spectrum of the sun, dark lines which were later studied in detail by Frauenhofer. He observed about 600 lines in the sun's spectrum and named the most intensive of them by the letters from A to H. In 1820 Brewster explained that these lines originate from the absorption processes in the sun's atmosphere. Similar observations were made by several researchers in the spectra of stars, flames, and sparks. In 1834 Wheatstone observed that the spectra produced with a spark depended on the electrode material used. Ångstrom in turn made the observation that spark spectra were also dependent on the gas surrounding the electrodes. The study of flame spectra became much easier after the discovery of the Bunsen burner in 1856.

Kirchhoff and Bunsen constructed a flame spectroscope in 1859. This new instrument made it possible to study small concentrations of elements which was impossible by the other methods available at that time. They also showed that the lines in the flame spectra originated from the elements and not from the compounds. Applications for this new technique were soon observed in astronomy and analytical chemistry. In the next five years, four new elements (Rb, Cs, Tl, and In) were found by flame emission spectroscopy.

The first quantitative analysis based on the flame emission technique was made by Champion, Pellet, and Grenier in 1873. They determined sodium by using two flames. One flame was concentrated with sodium chloride and the other was fed with the sample solution along a platinum wire. The determination was based on the comparison of the intensities of the flames by dimming the brighter flame with a blue glass wedge.

Diffraction gratings were studied by many scientists in the nineteenth century. By the end of the century gratings were improved markedly thanks to Rowland's studies. In the Rowland spectrograph the slit, grating, and camera were all in the same circle (Rowlands circle).

The main points to note of the spectroscopy of the nineteenth century were:

(i) By sufficient heating the monoatomic gases emit radiation spectra which consist of separate emission lines. Emission spectra of polyatomic gases consists of a number of lines close to each other, while solids and dense gases emit continuum radiation;

(ii) A cool gas absorbs radiation at the same frequencies as it emits radiation. If a continuum emission is directed into a cool gas vapour, the spectrum recorded will contain dark absorption lines or bands;

(iii) The frequencies (or wavelengths) of the lines are characteristic of each atom and molecule, and the intensities are dependent on the concentration.

Both the qualitative (the wavelengths of the lines) and the quantitative analysis (the intensities of the lines) are based on these phenomena.

Emission spectra were first utilized in analytical chemistry as they were simpler to detect than absorption spectra. Flames, arcs, and sparks are all classical radiation sources. Lundegårdh first applied a pneumatic nebulizer and an air-acetylene flame. The development of prism and grating instruments was parallel. Photography was employed to detect the spectral lines. The first commercial flame photometers came on the market in 1937.

The wavelength calibration using the red cadmium line at 643.8 nm was performed in 1907. Later the calibration was performed according to the green mercury line at 546.0 nm. In 1960 the new definition for the length of one metre was confirmed to be the wavelength of the krypton-86 line at 605.8 nm multiplied by the factor of 1650763.73. Nowadays one metre is defined as the distance which the light propagates in 1/299792458 seconds in a vacuum.

The basic concepts of atomic absorption spectrometry were published first by Walsh in 1955, this can be regarded as the actual birth year of the technique. At the same time Alkemade and Milatz designed an atomic absorption spectrometer in which flames were employed both as a radiation source and an atomizer. The commercial manufacture of atomic absorption instruments, however, did not start until ten years later. Since then the development of atomic absorption spectrometry has been very fast, and atomic absorption (AA) instruments very quickly became common. The inventions of dinitrogen oxide as oxidant and electrothermal atomization methods have both significantly expanded the utilization field of atomic absorption spectrometry. These techniques increased the number of measurable elements and lowered detection limits. Todays' graphite furnace technique is based on the studies of King at the beginning of the twentieth century.

The use of atomic emission spectrometry expanded markedly when the first commercial plasma atomic emission spectrometers came on the market in the middle of the seventies. The principle of the direct current plasma (DCP) source was reported in the twenties and the first DCP instrument was constructed at the end of the fifties. The first microwave plasma source was introduced in 1950, and the first inductively coupled plasma source was patented in 1963.

Atomic fluorescence in flames was first studied by Nichols and Howes. They reported the fluorescence of Ca, Sr, Ba, Li, and Na in a Bunsen flame in 1923. The first analytical atomic fluorescence spectrometer based on the studies of Bagder and Alkemade, was constructed by Winefordner and his co-workers in 1964.

Alan Gray first suggested the connection of a plasma source and a mass spectrometer in 1975. The direct current plasma jet was first applied in this new technique. Later it was shown that the inductively coupled plasma (ICP) met the requirements better than the DCP for an ionization source of mass spectroscopic analysis. The pioneering work of ICP-MS was mainly conducted by three research groups (Fassel, Gray, and Date).

Table 1 summarizes important steps of the history of atomic absorption and plasma emission spectrometry.

1.2 THE PRESENT STATUS OF ATOMIC SPECTROMETRIC METHODS

Atomic absorption, plasma atomic emission, and atomic fluorescence spectrometry are all optical atomic spectrometric techniques developed rapidly during the past years. These methods are based on the measurement of absorption, emission, or fluorescence originated from the free, unionized atoms or atomic ions in gas phase. The different instrument

Table 1   *Important steps in the history of atomic absorption (AAS), plasma atomic emission (plasma AES), atomic fluorescence (AFS), and plasma mass spectrometry (plasma MS)*

Year	AAS	Plasma AES, AFS, plasma MS
1916	Description of the hollow cathode lamp (Paschen)	
1922		The principle of the direct current plasma source (Gerdien and Lotz)
1928	Premix air-acetylene flame, pneumatic nebulizer and spray chamber (Lundegårdh)	
1941	Determination of Hg in air by AAS (Ballard and Thornton)	
1950		The first microwave plasma source (Cobine and Wilbur)
1953	The first patent in AAS (Walsh)	
1955	The principle of atomic absorption spectrometry (Walsh; Alkemade and Milatz)	
1958	First applications (Allan and David)	
1959	Graphite furnace (L'vov)	Analytical DCP source (Margoshes and Schribner; Korolev and Vainstein)
1961	The first book on AAS (Elwell and Gidley)	
1962	First commercial AAS instruments	
1963		The first analytical ICP source (Greenfield *et al.*)
1964		The first analytical atomic fluorescence spectrometer (Winefordner *et al.*)
1965	Dinitrogen oxide-acetylene flame (Amos and Willis) D_2-background correction (Koityohann and Pickett)	
1967	Graphite furnace (Massmann) Cold vapour method (Bradenberger and Bader)	
1969	Hydride generation methods (Holak)	
1970	Delves Cup method (Delves)	First commercial plasma AES instruments
1971	Zeeman-based background correction (Hadeishi and McLaughlin)	
1975	Matrix modification (Ediger)	DCP-MS (Gray)

Table 1 *Continued*

Year	AAS	Plasma AES, AFS, plasma MS
1977	Commercial constant temperature graphite furnace	
1978	Platform atomization and probe atomization (L'vov)	
1980		Introduction of IR lasers (1064 nm)
1983	Smith-Hieftje background Correction	First commercial ICP-MS instrument
1984	Stabilized temperature platform furnace	
1990	Horizontally heated graphite furnace	
1995		First commercial axial plasma ICP-AES instrument
1995		First commercial Time of Flight ICP-MS instrument
1997		Collision cell in commercial ICP-MS instruments
1999		Introduction a short wavelength UV Laser (213 nm)

components and their properties used in these techniques are often quite much alike and are related to the wavelength area covered. To simplify, similar detectors and monochromators are used. Samples are introduced usually as liquids through solution nebulization.

The "heart" of each measurement system is the atomizer/excitation unit (Figure 1). In atomic absorption and fluorescence spectrometry the atomizer should produce a large population of sample atoms that are in the ground state. The sample matrix should not affect the degree of atomisation, or otherwise chemical interferences are encountered. Different combustion flames are most commonly used as atomization devices. In addition, graphite furnaces and flame or electrically heated quartz tubes are commonly used to enhance the determination of low analyte levels by AAS. In AES the atomizer/excitation source must be very efficient or otherwise only the elements having low lying excited states can be measured with reasonable sensitivity (Figure 2). Various plasma sources, especially ICPs, fulfil this well, especially when liquid samples are used. The atomization/excitation devices used in AAS, AES and AFS techniques can be at the atomization temperature continuously (flame, plasma) or transiently (graphite furnace). Samples can be introduced continuously (solution nebulisation, continuous hydride

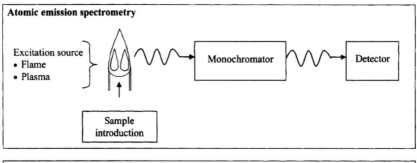

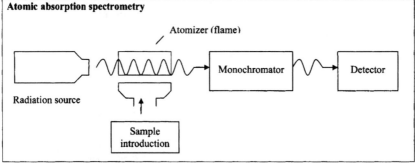

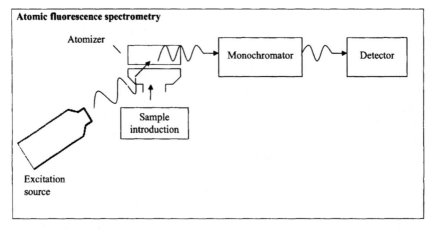

Figure 1   *Basic measurement arrangements used in different atomic spectrometric techniques*

generation), or transiently (flow injection technique, sample droplet introduced into a graphite furnace).

In atomic absorption spectrometry the measurement arrangement is linear. The narrow spectral band of electromagnetic radiation is passed through the atomic cloud and the specific atomic absorption is measured. Modern atomic absorption instruments are still in principle similar to the instruments from the early days of atomic absorption.

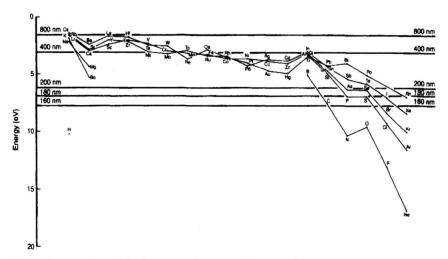

Figure 2   *Energies of the first excited states of the neutral atoms*
(Adapted from: D. A. McGregor, K. B. Cull, J. M. Gehlhausen, A. S.
Viscomi, M. Wu, L. Zhang and J. W. Carnahan, Anal. Chem., 1988, **60**,
1089A)

The most significant development has occurred in electronics. Micro-
processors have markedly simplified working with these instruments.
Modern instruments are faster and safer, and the performance with
respect to precision and accuracy has improved. The use of an
autosampler makes it possible to determine 6 elements in 50 samples in
35 minutes, *i.e.* about 500 determinations in one hour.

The wide popularity of AAS can be attributed in its simple and convenient use with various methods. In addition, the high sensitivity of
graphite furnace AAS is very important in many applications. However, the simultaneous multi-element analysis or qualitative analysis by
AAS is an arduous task. These two shortcomings of AAS (as well as
many advantages of AAS) derive from the line-like radiation source and
the atomizer employed. Because a line-like radiation source (a hollow
cathode lamp or an EDL) emits an extremely narrow radiation line that
is locked on the resonance lines of the atoms of the analyte element, it
provides relatively linear calibration graphs, minimizes spectral interferences, and makes the alignment of the instrumentation and selection of
wavelengths easy. However, use of hollow cathode lamps or EDLs, only
allows 1 to 3 elements to be determined at the same time, which
decreases the rate of multi-element analysis and makes qualitative
analysis impractical.

With few exceptions, graphite furnaces and flame atomizers are
both limited to use with liquid samples and are capable of effectively

atomizing only a fraction of the elements. Graphite furnace determinations require optimization of instrumental conditions for each element (temperature programme, observation time) in order to obtain optimal results. Thus, multi-element analysis is compromised. In addition, the GF-AAS techniques suffer from inter-element interferences and background absorption which must be overcome.

In order to make the lamp change rapid, various arrangements are offered by the manufacturers. For instance, by a ferris-wheel like turret a sequence of elements can be measured during each turret rotation. Another approach is a combination of a continuum radiation source and a high-resolution spectrometer. However, this combination has not achieved great acceptance. A common problem with continuum sources is their relatively low intensity in the UV region.

Possibilities in continuum AAS include the use of a Fourier transform spectrometer, television-like detectors with an echelle monochromator, a resonance monochromator, and an instrument based on resonance schlieren (Hook) spectrometry.

Before atomic absorption, atomic emission was used as an analytical method. The intensity of the emitted radiation and the number of emission lines are dependent on the temperature of the radiation source used. A flame is the oldest emission source. It is uncomplicated and its running costs are low. Flame emission is used, especially, for the determination of alkali and alkaline earth metals in clinical samples. Arcs and sparks are suitable radiation sources for multi-channel instruments in laboratories where several elements must be determined in the same matrix at high frequency, like in metallurgical laboratories. However, they are not readily applied for the analysis of liquid samples.

Various plasmas (ICPs, DCPs, MWPs) possess a number of desirable analytical features that make them remarkably useful multi-element atomization-excitation sources. This applies particularly to inductively coupled plasmas. The sample particles experience a gas temperature of about 7000 to 8000 K when they pass through the ICP, and by the time the sample decomposition products reach the analytical observation zone, they have had a residence time of about 2 ms at temperatures ranging downwards from about 8000 to 5000 K. Both the residence time and temperatures experienced by the sample are approximately twice as large as those in a dinitrogen oxide-acetylene flame. ICPs are therefore the most widely used plasma sources.

Plasma AES has several advantages (easy introduction of liquid samples, possibility for the qualitative and simultaneous multi-element analysis, measurements in the vacuum UV region, high sensitivity, low detection limits, less chemical interferences, low running costs) and it has become more and more important for the determination of traces in a great variety of samples. On the other hand, it does not compensate totally for any other instrumental method of analysis, but it compensates for those faults which might exist in other techniques. The complementary nature of plasma AES and AAS capabilities for trace elemental analysis is an important feature of these techniques. Plasma AES exhibits excellent power of detection for a number of elements which cannot be determined or are difficult to determine at trace levels by flame AAS *(e.g.* B, P, S, W, U, Zr, La, V, Ti) or by electrothermal AAS (B, S, W, U). Thus, optical plasma emission and atomic absorption are not actually alternatives, but in an ideal way complement one another. Currently it is possible to measure almost all elements in the Periodic Table, except some non metals with ICP-AES.

ICP is a very powerful excitation source. In fact most elements are to a large extent ionised in the normal observation zone of the ICP source (Figure 3). Hence the emission lines used are very often the lines of singly ionized atoms. An ICP is also suited very well for an ion source in mass spectrometry that allows easy introduction of liquid samples. Since the commercial introduction of ICP-MS instruments (VG Elemental Ltd.) in 1983, approximately 150 of them were installed worldwide during the first five years. Currently about ten companies manufacture ICP-MS instruments and thousands of instruments are in use worldwide.

Currently ICP-MS is being used in many branches of science. Many desirable analytical characteristics, such as superior detection limits,

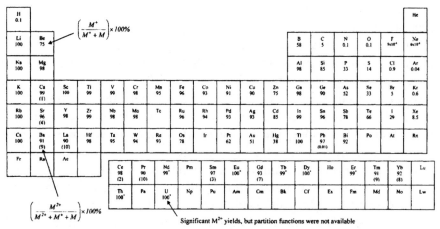

Figure 3 *The degree of ionization of various elements at ionization temperature 7500 K. Electron number density $n_e = 1 \cdot 10^{15}$ cm^{-3} Calculations are based on Saha equation*
(Adapted from: R. S. Houk, Anal. Chem., 1986, **58**, 97A)

spectral simplicity, possibility for simultaneous multi-element analysis, and isotope ratio determinations, are reasons for its widespread popularity. However, not even this technique is free from interferences. Particularly, spectral (polyatomic) and non-spectral (suppression and enhancement) interferences cause analysts to consider carefully the sample preparation procedure and finally the matrix. The development of laser ablation sample introduction has made ICP-MS a very powerful tool for the direct analysis of many types of solid samples.

Atomic fluorescence has many superior features for trace elemental analysis (spectral simplicity, wide dynamic range, and simultaneous multi-element analysis). In atomic fluorescence spectrometry the sample atoms in the atomizer are irrariated by an intense electromagnetic radiation from the source. Sample atoms are excited if the energy (wavelength) of the incoming radiation exactly matches with the energy states of the atoms in the atomizer unit. The excited atoms rapidly relax to ground state through collisions and emission of fluorescence radiation. The intensity of the fluorescence radiation is measured. The measured signal must not contain the radiation from the excitation source. Therefore the detector and the excitation source are usually positioned at a right angle with each other. The monochromator shown in Figure 1 is not always necessary in atomic fluorescence spectrometry, *i.e.* when an excitation source emits only the wavelengths specific to the atoms that are measured. Hence only these atoms are excited.

However, major practical problems of this technique are connected with the radiation source. Among various radiation sources lasers best meet the requirements for AFS. Atomic fluorescence has not become such a popular technique as plasma atomic emission or plasma mass spectrometry. The analytical applications of AFS have suffered from the lack of commercial instruments. The only commercial atomic fluorescence spectrometer is the Baird Plasma AFS system (not in market anymore) which consists of pulsed hollow cathode lamps for excitation and an ICP as an atomization cell. However, atomic fluorescence detectors are currently widely used in the determination of very low levels of mercury and the hydride forming elements (e.g. As and Se).

Table 2 describes the amount of published fundamental research papers in atomic spectrometry during the past few years. A brief survey of the current literature shows that each year fewer papers are dealing with novel AAS instrumentation, techniques, or applications. Instead, most studies are now concerned with plasma emission techniques (especially ICP-AES) or ICP mass spectrometry. Although the number of recent AAS papers is declining, the number of AAS determinations

Table 2 *Number of fundamental research articles[a] cited under some subtitles in "Atomic spectrometry update" (Source: Journal of Analytical Atomic Spectrometry)*

Topic	Number of articles cited[b]				
	2000	2001	2002	2003	2004
Atomic absorption, emission and fluorescence spectrometry					
Sample introduction					
Flow injection and preconcentration	89	163	103	85	196
Chemical vapour generation	115	96	50	82	136
Nebulization	37	25	10	13	19
Solid sampling	63	12	11	10	3
Electrothermal vaporization	15	13	9	8	3
Instrumentation	207	102	14	11	10
Fundamental studies	52	72	48	16	36
Laser-based analytical atomic spectrometry	68	110	122	49	77
Chemometrics	23	17	9	15	15
Coupled techniques for speciation	88	105	65	41	42
Number of articles included in update	676	722	443	330	539
ICP-MS					
Fundamental studies and instrumentation	10	20	20	19	7
Sample introduction, altogether	44	39	53	40	27
Laser ablation	22	16	27	30	11
Separation and speciation	12	12	21	18	11
Isotope ratio measurement	6	11	22	20	21
Number of articles included in update under ICP-MS	109	93	138	115	84

[a]Numerous applications of the different techniques are not included.
[b]Year of update of publication.

performed each year remains substantial, and the sales of AA instruments remain strong. This is clearly indicating that atomic absorption spectrometry has become a mature analytical technique during its existence of approximately 40 years.

Figure 4 presents the elements which can be determined by AAS and plasma AES methods.

1.3 TERMS AND DEFINITIONS

A newly developed analytical technique gives rise to new terminology. Existing terms may acquire a specialized meaning and completely new terms have to be invented. In order for people, working with atomic spectroscopy, to communicate with complete understanding several agreements and suggestions have been made by international bodies. The following nomenclatures are dealing with the spectroscopic terms and definitions: IUPAC (International Union of Pure and

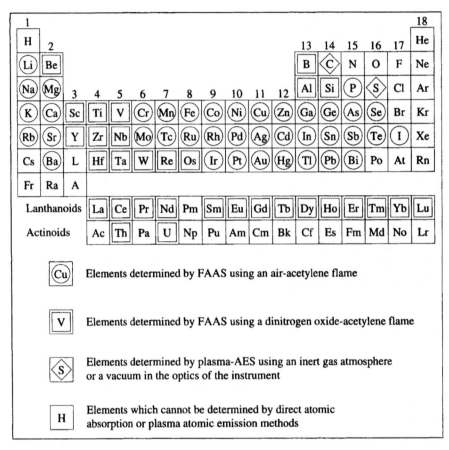

Figure 4 *Elements which can be determined by AAS and plasma AES methods*

Applied Chemistry), 'Nomenclature, Symbols, Units, and their Usage in Spectrochemical Analysis', Parts I and II, Pergamon Press, Oxford, 1975; IUPAC, 'Compendium of Analytical Nomenclature', 2nd Edition, Blackwell Scientific Publications Ltd, Oxford, 1987; R.C. Denney, 'A Dictionary of Spectroscopy', 2nd Volume, MacMillan Press, 1982. In the following text some common terms, definitions, and symbols associated with atomic spectrometry are given.

1.3.1 General Terms

Atomic absorption spectrometry (AAS). An analytical method for the determination of elements in small quantities. It is based on the absorption of radiation energy by free atoms.

Atomic fluorescence spectrometry (AFS). An analytical method for the determination of elements in small quantities. It is based on the emission of free atoms when the excitation is performed by the radiation energy.

Atomic emission spectrometry (AES). An analytical method for the determination of elements in small quantities. It is based on spontaneous emission of free atoms or ions when the excitation is performed by thermal or electric energy.

Molecular absorption spectrometry with electrothermal vaporization (ETV-MAS). An analytical method for the determination of elements in small quantities. It is based on the absorption of radiation energy by two atomic molecules at elevated temperatures.

Detection limit (DL). The minimum concentration or an amount which can be detected by the analytical method with a given certainty. According to the IUPAC recommendation, DL is the mean value of the blank plus three times its standard deviation.

Instrumental detection limit gives the smallest possible concentration which can be achieved by the instrument. The instrumental detection limit is derived by using the optimum instrumental parameters and the pure solvent (water) as a sample. The instrumental detection limit is useful in comparison of the performance of the different spectrometers.

The practical detection limit (PDL) is the smallest concentration of the analyte which can be obtained in a real sample. The PDL value may be 5 to 100 times greater than the corresponding instrumental detection limit.

Accuracy. The difference between the measured and the real concentration.

Precision. Can be obtained as the relative standard deviation (RSD) when the analysis is repeated several times under identical conditions.

Repeatability. Defined by the standard deviation of the results obtained in the same laboratory within a short time period.

Reproducibility. Defined by the standard deviation of the results obtained by different laboratories.

Sensitivity. The method is sensitive if small changes of the analyte concentration *(c)* or amount *(q)* affect great changes in the property (x). Thus, the ratio dx/dc or dx/dq is large and the sensitivity S of the method for an element i can be defined as the slope of the calibration graph.

Characteristic concentration and mass. The characteristic concentration of an element in atomic absorption is defined as its concentration or amount which gives a change, compared with pure solvent, of 0.0044 absorbance units (1% absorption) in optical transmission at the wavelength of the absorption line used. The characteristic mass is the mass of an element in picograms that gives 0.0044 integrated absorbance.

1.3.2 Terms Concerned with the Analytical Technique and Spectral Radiation

Characteristic radiation. Radiation which is specifically emitted or absorbed by free atoms of the given element.

Resonance line. The emission of an atomic line is the result of a transition of an atom from a state of higher excitation to a state of lower excitation. When the lower state of excitation is the ground state, the line is called the resonance line.

Hollow cathode lamp. A discharge lamp with a hollow cathode used in atomic spectrometry to produce characteristic radiation of the elements to be studied. The cathode is usually cylindrical and made from the analyte element or contains some of it.

Electrodeless discharge lamp (EDL). This is a tube which contains the element to be measured in a readily vaporized form (often as iodides). A discharge is produced in the vapour by microwave or radio frequency induction. The lamp emits very intensive characteristic radiation of the analyte.

Plasma. Partly ionized gas (often argon) which contains particles of various types (electrons, atoms, ions, and molecules) maintained by an external field. As a whole, it is electrically neutral.

Inductively coupled plasma (ICP). A spectroscopic source in which plasma is maintained by a magnetic field.

Direct current plasma (DCP). A spectroscopic source in which plasma is maintained by an electric field (a direct current arc between three electrodes).

Microwave plasma (MWP). A spectroscopic source where plasma is maintained by a microwave field.

Plasmatorch. An inductively coupled plasma source.

Plasmajet. A direct current plasma source.

Self absorption. The radiation emitted by the atoms of a given element is absorbed by the atoms of the same element in a spectral source.

Analyte. The element to be determined.

Matrix. The chemical environment of the element to be determined.

Matrix effect. An interference caused by the difference between the sample and the standards.

Sample solution. A solution made up from the test portion of the sample for the analysis.

Standard. A solution containing a known concentration of the analyte in the solvent with the addition of reagents used for preparation of the sample solution and other major constituents in proportions similar to those in the sample.

Blank. A blank test solution containing all the chemicals in the same concentrations as required for the preparation of the sample solutions except the analyte.

Spectrochemical buffer. A substance which is part of the sample or which is added to the sample, and which reduces interference effects.

Ionization buffer. A spectroscopic buffer which is used to minimize or stabilize the ionization of free atoms of the element to be determined.

Matrix modification. The alteration of the thermal pretreatment properties of the analyte or matrix by chemical additions.

L'vov platform. A small graphite platform inside the graphite tube on which the sample is deposited.

Slotted tube atom trap (STAT). A double slotted quartz tube supported above the air-acetylene flame of a conventional burner. One of the slots is set directly above the flame, and the tube is aligned with the optical path of the spectrometer.

Direct method. The method is direct if the atomic absorption, atomic emission, or atomic fluorescence of the analyte is related to its concentration.

Indirect method. Indirect atomic spectrometric method based on the chemical reaction of the analyte with one or more species, from which one should be measurable by AAS, AES, or AFS.

Flow injection analysis (FIA). FIA is a technique for the manipulation of the sample and reagent streams in instrumental analysis.

Hydride generation technique. Hydride generation is an analytical technique to separate volatile hydride-forming elements from the main sample matrix before their introduction into the light path of the instrument, and to convert them into an atomic vapour once they are there.

Cold vapour technique. An analytical technique for the determination of mercury. Mercury is first reduced to the metallic element, vaporized, and introduced into the light path of the instrument.

Zeeman effect. The splitting of spectral lines in a strong magnetic field.

Parts per million (p.p.m.). Milligrams of analyte per kilograms of sample (mg kg^{-1}).

Parts per billion (p.p.b.). Micrograms of analyte per kilograms of sample (μg kg^{-1}).

Part per trillion (p.p.t). Nanograms of analyte per kilograms of sample (ng kg^{-1}).

CHAPTER 2

Overview of Atomic Spectrometric Techniques

2.1 THEORY OF ATOMIC SPECTROSCOPY

2.1.1 Atomic Absorption, Emission, and Fluorescence Spectra

Each line in the line spectra can be considered as 'monochromatic' radiation. Because of the wave character of light, each line in the spectrum is characterized in terms of its wavelength (λ):

$$\lambda = c/v \tag{1}$$

where c is the velocity of light in a vacuum (about 3×10^8 ms^{-1}) and v is the frequency in Hertz (Hz) or cycles s^{-1}. Spectra can also be presented as a function of the wavenumber ($\bar{v}$):

$$\bar{v} = 1/\lambda \text{ (unit cm}^{-1}) \tag{2}$$

The recommended unit of wavelength is the nanometer (1 nm = 10^{-9} m). The traditional unit is the Ångstrom (1 nm = 10 Å). Other widely used units are the micron (1 μm = 10^{-6} m) and the millimicron (1 mμ = 1 nm = 10^{-9} m). The visible region of the spectrum extends from about 770 nm (red) to 380 nm (violet). However, wavelengths of adjacent spectral regions are overlapping, and in atomic spectrometry we shall mainly be concentrating on the spectral region extending from about 190 nm in the ultraviolet to about 850 nm in the infrared (Figure 5).

The distribution of the spectral lines of each individual element is not random. It was discovered first empirically and also later shown theoretically that the wavelengths of the lines of the simple atomic

16

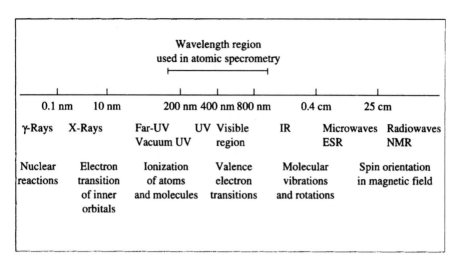

Figure 5 *Spectral regions of electromagnetic spectrum*

spectra can be fitted to simple series formulae with great accuracy. Furthermore, many of the lines in the simple spectra occur in small groups which are called multiplets, such as doublets of the alkali metals or triplets of the alkaline earths. There is also a constant difference between the wavenumbers of the two components of some doublets or two of the three components of some triplets. For example, the two lines of each doublet are separated by 17 cm^{-1} in the atomic spectrum of sodium (Table 3). This has been shown by Ritz to be a direct consequence of a general rule named the combination principle. According to this principle, for each atom or molecule there is a set of spectral terms which are such that the wavenumber of any line of the spectrum is the difference between two of the states.

The number of spectral terms of any atom is always smaller than its number of spectral lines, and the same spectral term appears in more

Table 3 *A part of the sodium doublets*

Wavelength(nm)	Wavenumber	Separation of Wavenumbers (cm^{-1})
819.48	12200	
818.35	12217	17
616.08	16227	
615.42	16244	17
568.82	17577	
568.26	17594	17
498.28	20064	
497.85	20081	17

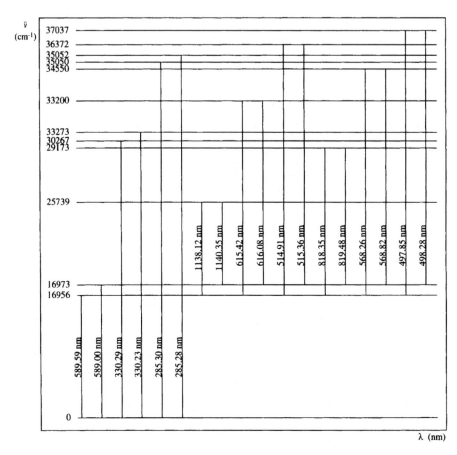

Figure 6   *Partial spectral term diagram for sodium*

than one state. Figure 6 shows the term diagram or level diagram of sodium. The horizontal lines represent spectral terms. The vertical lines show the spectral line between two terms. For convenience, the wavenumber of the lowest term is taken as zero. For example, the terms 16956 cm⁻¹ and 16973 cm⁻¹ both appear in seven different wavelengths.

An important advance in atomic spectroscopy was made in 1913 by Niels Bohr. According to the Bohr theory any atom is allowed only certain discrete and characteristic energy values. Absorption or emission of radiation is a result of a transition between two energy levels of the atom. These energy levels were found to be directly proportional to the empirical spectral terms of the atom. Planck showed the following relationship between the energy and frequency of a particular radiation:

$$E = hv = hc\bar{v} \tag{3}$$

where E is the energy and h is the Planck's constant (6.6×10^{-34} Js) The frequency of radiation (v) corresponds with the difference between two levels of energy E_m and E_n:

$$v_{mn} = (E_m - E_n)/h \qquad (4)$$

Thus, the experimental term diagrams become energy level diagrams of the atom by changing the scale.

The lowest energy level is called the ground state and is the state in which each atom will normally exist. Any other level of higher energy corresponds to an excited state. An atom can emit radiation only if it is in an excited state. Then energy will be released and the atom will return to a state of lower energy (either the ground state or an intermediate state). To get the emission spectrum of the atom it must be raised to an excited state by an external source of energy (flame, arc, spark, plasma, hollow cathode lamp, EDL). The absorption spectrum can be obtained when the atoms are irradiated with radiation of the correct wavelengths (corresponding to the energy difference between excited states and the ground state). Some of the energy will be absorbed to raise atoms to excited states. Only a few of all the possible transitions for a particular element will have the ground state as the lower of the two spectral terms, and hence the absorption spectrum is much simpler than the corresponding emission spectrum.

Bohr's model was based on the assumption that the energy states of an atom depends on the way in which the electrons move around the nucleus. Bohr proposed that the electrons move around the nucleus in circular orbits with different radii. This model was then extended by Sommerfeld, who postulated that the orbits could be elliptical. An electron can move from one orbit to another and at the same time the angular momentum changes in units of $h/2\pi$. The energy of the atom must then also change in multiples of a fixed unit, called a quantum or photon. The Bohr-Sommerfeld model only gives satisfactory results for hydrogen. The theory was also applied to other atoms during the period 1920 to 1930.

The modern atomic theory is based on quantum mechanics. The quantum theory was developed to explain the electronic structure of the atoms and the origin of the electromagnetic spectra. Optical spectra originate from the transitions of electrons in the outer orbitals. These electrons are called the valence or optical electrons. The outer orbitals of most of the atoms are incompletely filled, and the different energy states for a given atom can be obtained by calculating the various orbital possibilities of the optical electrons. Each possibility refers to a different, discrete state of the whole atom, and not only the energy of the electrons. The orbits of the electrons are described by a set of quantum numbers, which are used to estimate the number of spectral energy levels originating from given groups of valence electrons.

The energy states of an electron can be completely defined by a set of four quantum numbers (n, l, m_l, m_s). The principle quantum number of the orbital (n) determines the average distance of the electron from the nucleus, and it takes only positive integral values 1, 2, 3, *etc.* For each values of n there is a second quantum number l which takes the values of 0, 1, 2, . . . $n-1$. The different orbitals are named *s, p, d,* and *f* corresponding to the l values of 0, 1, 2, and 3, respectively. The magnetic quantum number m_l is connected to the orientation of the orbital in space and it ranges from $-l$ to $+l$. Every electron has a property called *spin* which is specified by an additional quantum number, m_s. Spin can be described as imagining the electron is a discrete particle like a top which is spinning about an axis passing through itself. Spin quantum numbers can be values $+\frac{1}{2}$ or $-\frac{1}{2}$ depending on whether the direction of spin is clockwise or counter-clockwise. Further, the rotation of electric charge about an axis gives rise to a magnetic moment that may point up or down depending on the direction of the spin.

Atomic orbitals (and electrons) are illustrated by the combination of the quantum numbers n and l (for example, $2s$, $3d$, $4f$). The way electrons occupy the orbitals in a given atom is called the electron configuration. In the hydrogen atom there is only one electron, and $1s$ denotes the ground state and $2s$, $3s$, $3p$, $3d$, . . . , various excited states. For many electron atoms the double occupancy of an orbital is denoted with a superscript 2. In addition, the Pauli exclusion principle (*i.e.* no more than two electrons per orbital) must be observed and each additional electron must be assigned to the orbital of lowest energy not yet filled. For example, the ground state configurations of the sodium and gallium atoms are $1s^2 2s^2 2p^6 3s$ and $1s^2 2s^2 2p^6 3s^2 3p^6 3d^{10} 4s^2 4p$, respectively. Now the superscripts indicate the total number of electrons in each set of similar orbitals. If there is one electron in the orbital, no superscript is used. Sodium and gallium have only one optical electron $3s$ and $4p$, respectively.

The state of the atom is defined by specifying its energy, orbital angular momentum, and spin (*viz*, its electron configuration). A state corresponds to the spectroscopic multiplet and consists of many components which have different values of the total angular momentum. The total angular momentum is a result of the combination of the orbital and spin angular momenta (Russell-Saunders or LS coupling scheme).

A single electron has the orbital angular momentum of $[l(l + 1)]^{1/2}$. For an atom with many electrons, the vector sum of l quantum numbers gives the total angular momentum L. Thus, the total orbital angular momentum of the atom is $[L(L + 1)]^{1/2}$. The symbol M_L represents a component of L in a reference direction (m_l for a single electron). Further, the total spin quantum number S is given by $[S(S + 1)]^{1/2}$, and it represents the electron spin angular momentum. The component of S in a reference direction is designed by M_S (m_s for a single electron).

The capital letters S, P, D, F, ... are used for the states with $L = 0, 1, 2, 3, \ldots$, respectively. Thus, these symbols are analogous to those for the orbitals of single electrons.

For a state with the total spin angular momentum $S = 1$, the spin multiplicity M_S is three (there are three M_S values: 1, 0, −1). The spin multiplicity is equal to $2S + 1$. The following examples should make the use of the symbols clear: for $M_L = 4$, $S = 1/2$, the symbol is ^2G; for $M_L = 2$, $S = 3/2$, the symbol is ^4D; for $M_L = 0$, $S = 1$, the symbol is ^3S. In speaking of states with spin multiplicities of 1, 2, 3, 4, ... we call them singlets, doublets, triplets, quartets, ..., respectively. Thus, the three example states above should be called doublet G, quartet D, and triplet S, respectively.

As in the case of a single electron, the total angular momentum of the entire atom (J) is obtained as the vector sum of L and S.

The various states of any atom are given usually in the form $n^a T_j$, in which n is the main quantum number or electron configuration (for example $3d^4D_2$, where $n = 3d^4$), a is $2S + 1$ or the multiplicity of the term, and T is S, P, D, F, G, ... corresponding with the L values of 0, 1, 2, 3, 4, ..., respectively, and j is the numerical value of J.

J values may be used for measuring intensities of spectral lines because the statistical weight of each state (g) is equal to $2J + 1$.

In atomic absorption spectroscopy, the spectral lines originating from neutral atoms are normally used, but in plasma emission spectral lines of ions may also be used. The ionization state of an element is usually indicated in spectroscopy as Roman numerals in the following way: M I: neutral atom (M^0); M II: unicharged ion (M^+); M III: dicharged ion (M^{2+}); *etc.*

2.1.2 Emission and Absorption of Energy

As described above, each atom has a number of possible energy levels or states. Emission or absorption of radiation originates from an electron transition between particular pairs of these states. Which transitions are possible, can be predicted by the spectroscopic selection rules based on quantum mechanics. The quantity of energy emitted or absorbed (the relative intensities of the spectral lines) can also be predicted by the quantum theory together with the positions of the spectral lines. A fixed number of atoms exist in each of the possible states for each element under the given conditions. The population of each level will depend on the amount of energy available to the atoms. Environmental factors (temperature and pressure) will have an affect on the intensities of the lines.

The atomic emission spectrum of sodium includes an intense line at 589 nm (doublet) and another line at 569 nm. Part of the energy levels for sodium are shown in Figure 7 together with the allowed emission (four) and absorption (two) transitions. In the ground state the lone optical electron occupies the $3s$ orbital. The transition of the electron to any higher energy level leads to the corresponding excited state. According to the selection rule, transitions between levels in which the quantum number l changes by one unit (*i.e.* $\Delta l = + 1$ or -1) are allowed. Thus, according to the theory transitions between two orbitals of the same type are forbidden. However, these transitions occur to some extent, but the intensities of the corresponding spectral lines are low.

Einstein's theory of radiation is a method of classifying and explaining the relative line intensities. According to the radiation theory the intensity of a line corresponding to a particular transition will be directly proportional to its transition probability. The calculations are based on the rules of quantum mechanics, are complex, and only possible when simplifying assumptions are made.

Einstein's theory involves three coefficients, and are defined for spontaneous emission (A_{ij}), absorption (B_{ji}) and stimulated emission (fluorescence) (B_{ij}) (Figure 8). A_{ij}, is the probability that an atom in state 'i' will spontaneously emit a quantum $h\nu$ and pass to the state 'j'. The unit for A_{ij} is s^{-1}. The other two coefficients are more difficult to define, since the probability of an absorption or fluorescence transition will depend on the amount of incident radiation. The transition probability in these cases is obtained by multiplication of the appropriate coefficient and the transition density at the frequency corresponding to the particular transition (ρ_ν). B_{ji} and B_{ij}, have the unit s^{-1}. The radiation density (ρ_ν) is defined as energy per unit volume, and it has the units erg cm^{-3} or g cm^{-1}s^{-1}. The unit of B_{ji} and B_{ij} is then cm g^{-1}.

Einstein has derived several useful relationships between the A_{ij}-, B_{ji}- and B$_{ij}$-coefficients by assuming thermodynamic equilibrium between the radiation and the atoms and comparing it with the equilibrium of a black-body radiator at the same temperature:

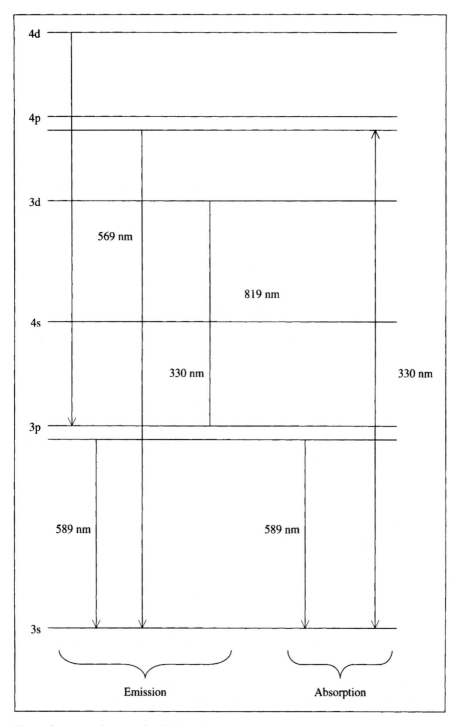

Figure 7 *Part of energy levels for sodium*

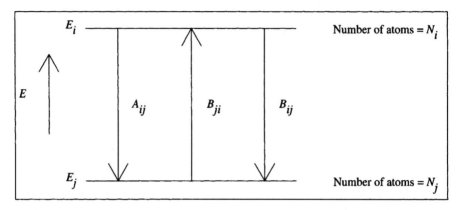

Figure 8   *Spontaneous emission (A_{ij}), absorption (B_{ji}), and fluorescence (B_{ij})*

$$A_{ij}/B_{ji} = 8\pi h v^3/c^3 \tag{5}$$

$$g_i B_{ij} = g_j B_{ji} \tag{6}$$

where g_i and g_j are the statistical weights of the states 'i' and 'j'.

A_{ij}, is directly proportional to the oscillator strength (f_{ij}), as shown by Landenburg's equation:

$$A_{ij} = 8\pi^2 e^2 f_{ij}/\lambda^2 mc \tag{7}$$

where e and m are the electron charge and mass, respectively. Substitution of the values of the constants will give:

$$A_{ij} = 0.6670 \times 10^{-14} f_{ij}/\lambda^2 \tag{8}$$

where the units of A_{ij} and λ are s^{-1} and nm, respectively. The absorption oscillator strength (f_{ij}) is directly proportional to the emission oscillator strength (f_{ij}):

$$g_j f_{ji} = g_i f_{ij} \tag{9}$$

Oscillator strength values are usually given as *gf*-values, which refer to emission or absorption lines corresponding to the transition between two particular states of the atom.

The number of transitions or photons per second for a spontaneous emission is $A_{ij}N_i$. The intensity of the corresponding emission line is:

$$I_{em} = h v A_{ij} N_i \tag{10}$$

For absorption and fluorescence, the numbers of transitions are $B_{ji}\rho_v N_j$ and $B_{ij}\rho_v N_i$, respectively. The total reduction in intensity will depend on the difference between these two numbers, and for absorption the intensity will be:

$$I_{ab} = hv(B_{ji}N_j - B_{ij}N_i)\,\rho_v \qquad (11)$$

On the other hand, for fluorescence the intensity will be:

$$I_{fl} = hvB_{ij}\rho_v N_i \qquad (12)$$

In practice, it is more convenient to use the incident intensity rather than radiation density:

$$I = (c/4\pi)\rho_v \qquad (13)$$

In this formula it is assumed that the incident light (intensity $= I$) is isotropic in the same way as the spontaneously emitted light (intensity $= I_{em}$). The incident light is in the form of a pencil. If the cone of light has an angle Ω, the intensity of the incident radiation beam is:

$$I_0 = (\Omega/\pi)I \qquad (14)$$

For a beam of unit solid angle this becomes $I/4\pi$, and the intensities of absorption and fluorescence will be

$$I_{ab} = (hv/c)(B_{ji}N_j - B_{ij}N_i)I_o \qquad (15)$$

$$I_{fl} = hvB_{ij}N_i I_o/c \qquad (16)$$

These three expressions (I_{em}, I_{ab}, and I_{fl}) form the practical basis of the formulae used in analytical atomic spectrometry.

2.1.3 The Maxwell-Boltzmann Law

The relative intensities of the spectral lines depend on the relative populations of the atomic states. In a thermal equilibrium the relative populations of the states of the given atom can be obtained by the Maxwell-Boltzmann law. If at an absolute temperature, T there are N_i atoms with an energy E_i, and N_j atoms with an energy of E_j, then the ratio of these atoms is (provided that both states are non-degenerate):

$$N_i / N_j = e^{-(E_i - E_j)/kT} \qquad (17)$$

where k is the Boltzmann constant. The number of particles with energy E_i is $(2J + 1)$ n_i (n_i is the population of each of the non-degenerate sublevels). Hence, $N_i = g_i n_i$. Since the law should be applied to non-degenerate levels of energy, E_i and E_j, it can be written as:

$$N_i / N_j = (g_i / g_j)(n_i / n_j) = (g_i / g_j)e^{-(E_i - E_j)/kT} \qquad (18)$$

Because absorption is a result of an electron transition from the lower level (ground state) to the higher level (excited state), the amount of absorption is dependent on the number of atoms in the ground state. According to the Maxwell-Boltzmann law the number of excited atoms will increase with increasing temperature. The wavelength of the line is inversely proportional to its excitation energy, and the number of excited atoms will increase exponentially with the increasing wavelength.

Figure 9 shows N_i/N_j ratios for some elements at varying temperatures. Because in AAS the atomization temperature lies generally below 3000 K and the wavelengths used are usually less than 600 nm, the amount of excited atoms is extremely small and can be neglected.

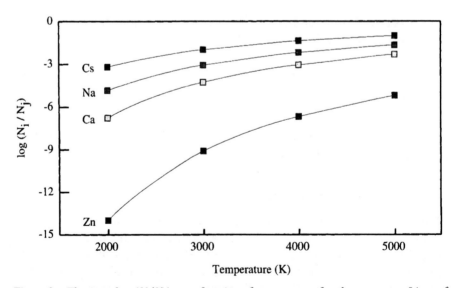

Figure 9 *The ratio log (N_i/N_j) as a function of temperature for the resonance Lines of Cs (852.1 nm; $g_i/g_j = 2$), Na (589.0 nm; $g_i/g_j = 2$), Ca (422.7 nm; $g_i/g_j = 3$) and Zn (213.8 nm; $g_i/g_j = 3$)*
(The values are obtained from R. J. Jaworowsky and R. P. Weberling, At. Absorpt. Newl., 1966, **5**, 125)

AAS measurements are based on Beer's law, and the ratio of radiation absorbed in the sample and radiation transmitted through the sample is measured. The absolute amount of absorbed radiation energy will depend on the wavelength, but the measurement procedure will eliminate the influence of the wavelength. In emission measurements the sensitivity will increase with increasing wavelength.

2.1.4 Spectral Linewidths

AAS is based on measuring the absorbance of a spectral line by using a radiation source emitting a sharp line. Even the sharpest line that can be produced by a modern spectrometer has a finite width which is extremely important in applications of both AAS and AFS. The profile of a spectral line is presented in Figure 10.

Figure 11 illustrates how sharp atomic lines are. In this figure the lead spectrum is plotted with the molecular absorption spectrum of ethanal. The lead spectrum is a sharp line, whereas ethanal has a broad absorption band with fine structure. This is a consequence of the different energy levels (Figure 12). Free atoms cannot by their nature have vibrational or rotational fine structures in their energy levels like the molecules do.

The width of any spectral line is defined in terms of its half-width ($\Delta\lambda_{1/2}$). A spectral line from any radiation source, or from absorption or emission cells, will be broadened by several factors. Some of them are

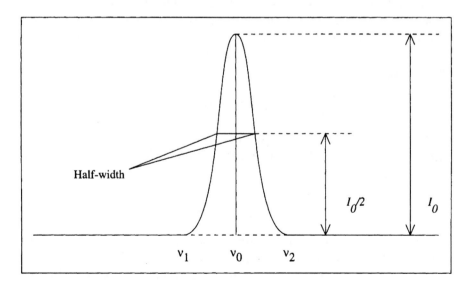

Figure 10 *The profile of a spectral line*

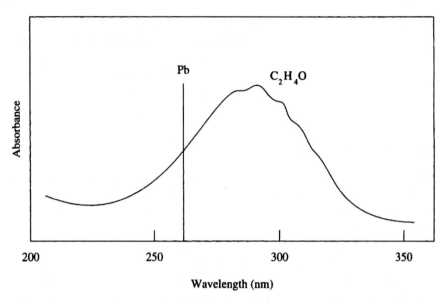

Figure 11   *The atomic absorption spectrum of lead and the molecular spectrum of ethanal*

more important than others. In addition, the contributions of different factors will vary with the source. The most common broadening effects are described below.

2.1.4.1   Natural Line Broadening.   Natural line broadening is the consequence of a finite lifetime of an atom in any excited state. The absorption process is very fast being about 10^{-15} s. The lifetime of the excited state is longer (about 10^{-9} s), but sufficiently short that the Heisenberg's Uncertainty Principle is appreciable. If the mean time the atom spends in an excited state, E_i is Δt_i, then there will be an uncertainty, ΔE_i in the value of E_i:

$$\Delta E_i \Delta t_i \geq h/2\pi \tag{19}$$

If the frequency for a transition between the levels E_i and E_j is $v_0 = \Delta E/h = (E_i - E_j)/h$ and the line has a finite width, Δv_N, it may be shown that:

$$\Delta v_N = (1/\Delta t_i + 1/\Delta t_j)/2\pi \tag{20}$$

This equation may be simplified for a resonance line. E_j will then be the ground state with an infinitely large Δt_j value. Δt_i is then proportional to the radiative lifetime of the excited state, which is defined for the resonance level as $1/A_{ij}$:

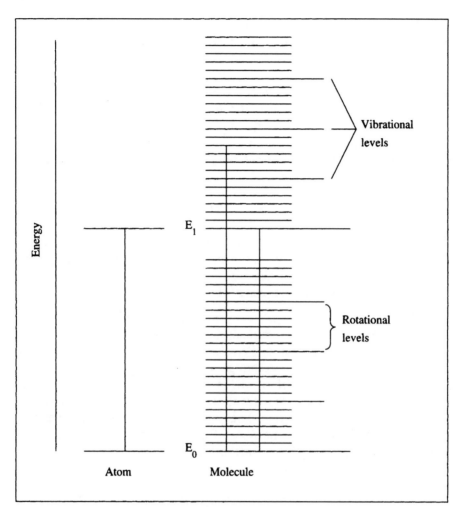

Figure 12 *Energy level diagram of an atom and a molecule*

$$\Delta v_N = A_{ij}/2\pi \qquad (21)$$

The natural broadening, Δv_N, is of the same order as A_{ij} (about $10^8 \, s^{-1}$). At the wavelength of 300 nm it will correspond to about 0.000005 nm, which is negligible in comparison with other line broadening factors.

2.1.4.2 Doppler Broadening. The Doppler broadening effect is caused by the thermal motion of the emitting or absorbing atoms. Under typical conditions the speeds of atoms are about $1000 \, m \, s^{-1}$. Although this is considerably smaller than the speed of light ($c = 3 \times 10^8 \, m \, s^{-1}$), it is fast enough for the Doppler effect to become noticeable.

If there is a relative speed, v, between the observer and the atom emitting light of frequency v, the light received by the observer would have the frequency of $(v + dv)$ or $(v - dv)$ depending on the direction of the motion $(dv = vv/c)$. Because in the gas phase atoms move randomly in all directions, the observer will receive light of many slightly different frequencies. As a consequence, the line will appear broadened. The distribution of the observed frequencies will be the same as that of the velocities of the atoms, and a Gaussian curve will represent the shape of the line. The line broadening due to the Doppler effect may be calculated from the formula:

$$\Delta v_D = v/c(2RT/M)^{1/2} \qquad (22)$$

Thus, the Doppler broadening is directly proportional to frequency of the line (v) and to the square root of the absolute temperature (T), and inversely proportional to the square root of the atomic mass (M). The Doppler effect accounts for most of the line width.

2.1.4.3 Pressure Broadening. Collision of an emitting or absorbing atom with other atoms or molecules affects broadening, shift, and asymmetry of the line profile. Pressure broadening is also known as collisional or Lorentzian broadening. According to the Lorentz theory, when an emitting or absorbing atom collides with a foreign particle there is an interruption of the vibration within the atom of an electron responsible for the spectral line. This is seen in line broadening which is directly proportional to the number of collisions per second per atom. The pressure broadening (Δv_P) for the resonance line may be calculated from the formula:

$$\Delta v_P = \sigma_L^2 N_2 [2\pi RT(1/M_1 + 1/M_2)]^{1/2}/\pi \qquad (23)$$

where σ_L^2 is the collision cross-section of the atom concerned, N_2 is the number of foreign particles per unit volume, R is the gas constant, and M_1 and M_2 are the relative mole masses of the colliding atom and foreign species. Equation 23 is useful to obtain the actual value of Δv_P, but it does not deal with the accompanying shift and asymmetry of the line. It has been shown that the frequency shift (Δv_S) depends on the Lorentzian line broadening (Δv_P) according to the following approximation:

$$\Delta v_S = 0.36\Delta v_P \qquad (24)$$

The relatively high concentration of foreign molecules in flame AAS results in a Lorentzian broadening line value similar to that of the

Doppler broadening value. The frequency shift may be considerable and lead to a reduction of the observed absorbance value at the central frequency (v_0) of the line, when a narrow line radiation source is used. In low pressure radiation sources (hollow cathode and electrodeless discharge lamps), which contain only inert gas atoms Δv_P and Δv_S will usually be negligible in comparison with Δv_D. For example, the collision cross-section for broadening of a calcium line at 422.673 nm by neon atoms has been found to be 7.4×10^{-15} cm^2. According to Equation 23, the Lorentz half width for a pressure of 1.33 kPa (Ne) is 0.00001 to 0.00003 nm depending on the temperature, whereas the Doppler broadening under the same conditions may vary between 0.0002 and 0.0008 nm.

If the interaction involves charged particles the pressure broadening is called the Stark effect, while collisions with uncharged particles is due to the van der Waals effect, and collisions between the same kind of atoms gives rise to the Holtzmark or resonance broadening effect.

The spectral line width is mainly determined by the Doppler and pressure line broadening effects, and the total line broadening (Δv_T) may be written as

$$\Delta v_T \approx \Delta v_P/2 + [(\Delta v_P/2)^2 + (\Delta v_D/2)^2]^{1/2} \qquad (25)$$

2.1.5 The Zeeman Effect

The Dutch physicist Pieter Zeeman discovered in 1896 the splitting of spectral lines in a strong magnetic field (field strength of several kilogauss). The observed line splitting is due to splitting of the energy levels (terms) in the atom. There are two types of Zeeman effects called (a) normal Zeeman effect, and (b) anomalous Zeeman effect.

In the case of the normal Zeeman effect, the emitted or absorbed line of an atom splits into three components under the influence of a magnetic field (Figure 13). The energy and wavelength of the central component (π) are unchanged with respect to the original values. The π-component is also linearly polarized to the magnetic field. The two other components are shifted by equal wavelength intervals higher (σ^+) and lower (σ^-) than the original wavelength. The sigma components are polarized perpendicular to the magnetic field. The extent of the shift depends on the applied magnetic flux density (magnetic field strength). The sum of the intensity of the three components is always equivalent to the intensity of the original, unaffected line. The distribution of the energy or intensity between the three components is $\sigma^+ : \pi : \sigma^- = 25:50:25$.

Splitting of the spectral lines into three components according to the normal Zeeman effect occurs only for atoms with singlet lines (terms

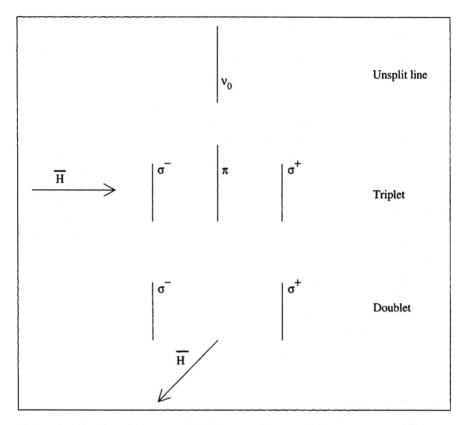

Figure 13   *The normal Zeeman effect splitting of a spectral line in a magnetic field*

with $S = 0$). Singlet lines are the main resonance lines of the alkaline earth metals (Be, Mg, Ca, Sr, Ba) and the Zinc group metals (Zn, Cd, Hg).

For the normal Zeeman effect the energy levels are expressed as:

$$E_{m_J} = \mu_B g B m_J \tag{26}$$

where μ_B is the Bohr magneton, g is the Landen factor, m_J is the projection of J parallel to the magnetic field, and B is the density of the magnetic flux. The Landen factor, g is written by the quantum numbers J, L, and S:

$$g = 1 + [J(J + 1) + S(S + 1) - L(L + 1)]/2J(J + 1) \tag{27}$$

In the case of the singlet term $g = 1$, because $J = L$. Thus, the difference between the energy levels is $\mu_B B$ and is independent of the symbol. This is visualized in Figure 14. According to the selection rule $\Delta m_J = 0, \pm 1$,

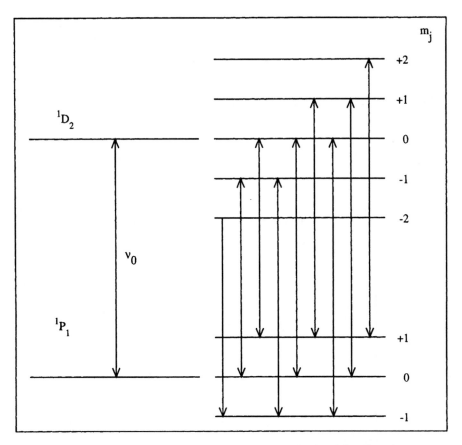

Figure 14 *The splitting of the singlet term in a magnetic field and the allowed transitions*

three spectral lines are observed perpendicularly to the magnetic field. The shift of the lines is expressed by the equation:

$$\pm\Delta v = eB/4\pi m_e \qquad (28)$$

where e is the charge of the electron and m_e its mass.

The mechanism of the anomalous Zeeman effect is exactly the same as that of the normal Zeeman effect, but it exhibits more than three components. For the anomalous Zeeman effect it is characteristic that the π-component also splits into several lines and thus no longer coincides exactly with the original resonance line. In this case the Landen factor, g will vary for various terms which is the reason for the splitting of the spectral lines into the several components (Figure 15).

Zeeman splitting patterns for a number of elements together with their resonance lines are shown in Figure 16.

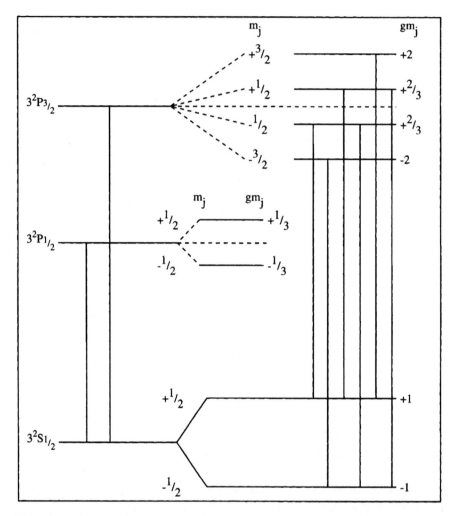

Figure 15 *The anomalous Zeeman effect splitting and the allowed transitions*

In strong magnetic fields the magnetic energy of an atom will be higher than the splitting of the lines. Then the *L-S* interaction will become negligible and the anomalous Zeeman effect resembles the Zeeman triplet (the Paschen – Back effect).

2.1.6   The Absorption Coefficient

Free atoms in the ground state can absorb radiation energy of exactly defined frequency *(hv)*. The absorption coefficient *(k_v)* at a discrete frequency is defined as:

$$I_v = I_v^0 e^{-k_v b} \tag{29}$$

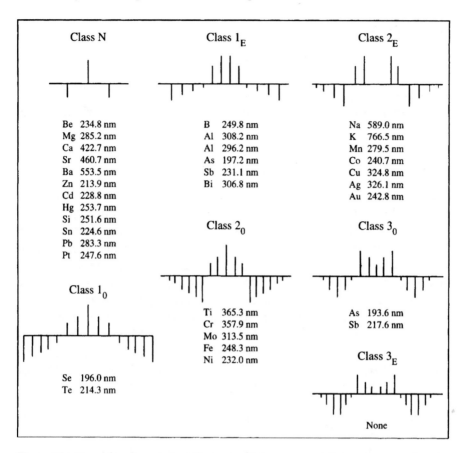

Figure 16 *Normal and anomalous Zeeman splitting patterns of some common elements*

where I_v^0 and I_v are the initial and final intensities of radiation at frequency v passed through an atom vapour of length b. The profile of the absorption line is obtained by plotting k_v as a function of frequency. The shape of the peak is the same as presented for the emission line in Figure 10. Integration of the absorption coefficient with respect to the frequency over the whole line will give the integrated absorption coefficient, K:

$$K = \int k_v dv = (hv_0/c)(B_{ji}N_j - B_{ij}N_i) \qquad (30)$$

where v_0 is the frequency at the centre of the line, and the small variation of v throughout the line is omitted. Because in the case of the absorption measurements $N_i \ll N_j$, equation 30 may be simplified as:

$$K = (hv_0/c)B_{ji}N_j \qquad (31)$$

The substitution for B_{ji} gives:

$$K = (\pi e^2/mc)f_{ji}N_j \tag{32}$$

According to equation 32, the amount of absorption is proportional to the oscillator strength (f_{ji}) and to the number of absorbing atoms (N_j).

2.1.7   Basic Concepts of the Plasma

Partly ionized gas or vapour is called a plasma. It contains atoms, molecules, and ions from which some fraction may be in excited states, and free electrons. Several theoretical models have been presented to describe a plasma. One of these is the so called Thermal Equilibrium Theory, which is based on the micro reversible principle. According to this principle, each energy process is in equilibrium with a reverse process. For example, the number of transitions per time unit from the state 'j' to the state 'i' (absorption) is exactly the same as the number of the reverse transitions (emission). According to Maxwell, the microscopic states of the plasma at thermal equilibrium may be calculated on the basis of the temperature, which is the only variable. The number of particles (dN) with the speed between $v \pm dv$ is:

$$dN = 4\pi N v^2 (M/2\pi kT)^{3/2} e^{-Mv^2/2kT} dv \tag{33}$$

where M is the relative mole mass of the particle, N is the number of particles, and k is the Boltzmann constant. At any thermodynamic equilibrium the population ratio N_i/N_j for the states 'j' and 'i' may be calculated from the Maxwell-Boltzmann equation (18).

The number of particles at a given state 'k' can be obtained using equation 34:

$$N_k/N = g_k/Z(T)e^{-E_k/kT} \tag{34}$$

where

$$Z(T) = \sum_j g_k e^{-E_k/kT} \tag{35}$$

Theoretically, there are an infinite number of states close to each other before the ionization energy. In practice, $Z(T)$ does not become infinite, because the electric field of close particles diminishes the ionization boundary and the number of energy levels becomes finite.

The ionization degree can be obtained at any temperature from Saha's equation:

$$K_I(T) = \frac{(2\pi m_e)^{3/2}(kT)^{3/2}}{h^3} \cdot \frac{2Z_I}{Z_A} \cdot e^{-E_I/kT} \qquad (36)$$

where $K_I(T)$ is the equilibrium constant, E_I the ionization energy, and Z_I and Z_A the partition functions of the ionized and atomic states, respectively. The intensity of the spectral line in plasma is defined by the Einstein theory (Equation 10).

A complete thermal equilibrium is not achievable using laboratory plasmas. The temperature is very high inside the plasma, but a few millimetres away from the plasma is the lower room temperature. Hence, plasma could be considered to be formed from a number of layers, and each layer is at a local thermal equilibrium. Secondly, most of the radiation is emitted to the environment without back absorption, which results in an increase of atoms in the ground state. However, the thermal equilibrium theory is a good approximation for relatively large particle concentrations.

The most common deviation from local thermal equilibrium is caused by the relative oversaturation of the ground state. The higher the energies of these levels, the more precisely the Boltzmann law gives the population ratio between two energy levels (N_i/N_j). If the deviations from the local equilibrium are great, the contribution of the electron density must be added to the temperature dependent Boltzmann law. Several plasma models have been proposed for these cases.

The emission intensity has a maximum with increasing temperature. With a plasma source it is possible to achieve higher temperatures than required from the analytical point of view. The intensity of the continuum emission of free electrons depends on electron number density (n_e) and electron temperature (T_e) according to Equation 37:

$$I(v) = Kn_e^2/T_e^{1/2} \qquad (37)$$

where K is a constant. This is why the hottest part of the plasma is not used for the analytical measurements.

2.2 INSTRUMENT COMPONENTS

As seen earlier (Figure 1), there are many similar instrument components/functioning in all atomic spectrometric techniques. However, the instrument components needed and their specifications depend on the particular technique used. Atomic absorption process, for example, is a very specific phenomenon. When the radiation source emits very narrow spectral lines (and also absorption lines are narrow), the resolution of the monochromator needs not to be very high, although the sample matrix is fairly complex. Otherwise, in atomic emission spectrometry employing plasma sources, very line rich spectrum is obtained. Consequently, a high resolution monochromator is needed to avoid spectral interferences. In this chapter the principles of different instrument components/functioning are overviewed. A more comprehensive discussion is given under chapters dealing with individual measurement techniques since each technique has it own specific demands for a particular instrumental feature, for example sample introduction.

The common functions in all atomic spectrometric techniques (and in ICP-MS) are sample introduction, atomization/ionization/excitation of the analyte, isolation of analytical signal and its detection (Table 4). Additionally, an external excitation source is needed when atomic absorption or fluorescence signal is measured.

Table 4 *General features of atomic absorption and atomic emission spectrometry and inductively coupled plasma mass spectrometry*

Feature	Technique		
	AAS	*AES*	*ICP-MS*
Sample introduction	Nebulization Vapour generation (hydride generation method, cold vapour method)	Nebulization Vapour generation Electrothermal vaporization Laser ablation	Nebulization Vapour generation Electrothermal vaporization Laser ablation
Atomization, excitation and ionization	Flame, graphite furnace, electrically heated quartz tube	Inductively coupled plasma, direct current plasma, microwave plasma	Inductively coupled plasma
Isolation of analytical signal	Monochromator	Monochromator (good resolution needed)	Mass analyzer (e.g. quadrupole, double-focusing instrument with electric and analyzers, magnetic sector time of flight spectrometer)
Measured property	Intensity ratio ($A=\log(I_0/I)$)	Emission intensity	Ion counting
Detection of analytical signal	Photomultiplier tube Charge coupled device Charge injection device	Photomultiplier tube Charge coupled device Charge injection device	Electron multiplier

2.2.1 Radiation Sources

An external radiation source is needed in atomic absorption and atomic fluorescence spectrometry. In principle, same kind of radiation sources can be used in both techniques. However, in AFS the intensity of the source should ideally be high to obtain good sensitivity in the measurement and therefore somewhat different sources are used in AFS (Chapter 7).

2.2.1.1 Hollow Cathode Lamps. Hollow cathode lamps are the most widely used radiation sources in the AA technique. A hollow cathode lamp consists of a glass cylinder, and an anode and a cathode (Figure 17). The cylindrical cathode is either made of the analyte element or filled with it. The diameter of the cathode is 3 to 5 mm. The anode is in the form of a thick wire and usually made of tungsten, nickel, tantalum, or zirconium. The glass tube is first evacuated and then filled with an inert gas (argon or neon). The pressure of the inert gas is about 0.5 to 1.3 kPa.

The choice of the inert gas depends upon two factors. Firstly, emission lines of the filler gas must not coincide with the resonance lines of the analyte element. The main emission regions from neon and argon are shown in Figure 18. The filler gas used in the hollow cathode lamp is easy to detect by the colour of the emission beam of the lamp. Argon gives a blue and neon an orange discharge. Secondly, the relative ionization potentials of the inert gas and cathode metal must be taken into account. The ionization potential of neon is much higher than that of argon. Neon is therefore used in lamps of metals with high ionization potentials. The kinetic energy of ions hitting the surface of the cathode must be higher than the energy of metal bonds. Normally, the emission intensity of the lamp can be increased by raising the applied current. This increases the number of metal atoms excited by filler gas ions in

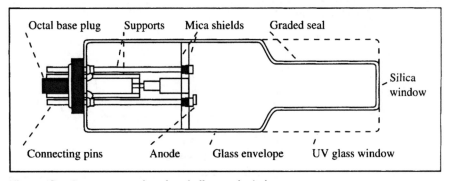

Figure 17 *Construction of modern hollow cathode lamp*
(Philips Scientific)

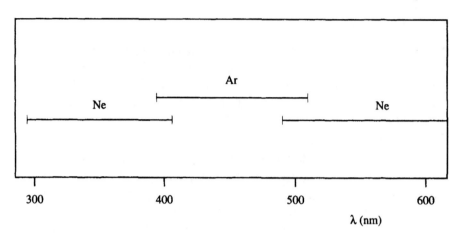

Figure 18   *Main emission regions of argon and neon*

the gas phase. In the case of easily volatilized metals, the increased current and resulting higher temperature lead to increased vaporization of the cathode material, which increases the number of collisions with neutral and charged particles. This is seen as a broadening (Doppler broadening) of the emission line. The cloud of neutral metal atoms in the ground state at the cathode opening can now absorb radiation at the emission maximum which is seen in the resulting distorted emission line profile. This phenomena is called self-absorption. At the position of the greatest probability of absorption there is a minimum on the emission line. The emission line profile can be optimized by adjustment of the applied current (Figure 19). Figure 20 shows the relation between the current and the pressure of the filler gas.

The operation of the hollow cathode lamp is as follows: (i) A voltage of 100 to 400 V is applied between the anode and cathode. Highly

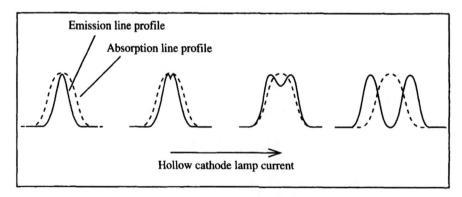

Figure 19   *The effect of the current on the shape of the emission line*

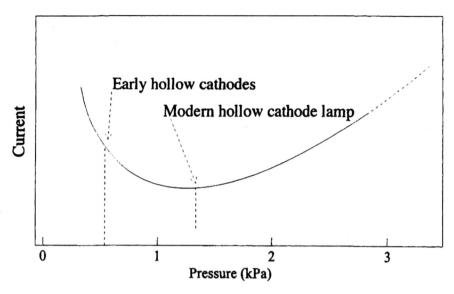

Figure 20 *Pressure/current characteristic of hollow cathode lamps*

energetic electrons emitted from the cathode ionize the filler gas as a result of collisions: $Ar + e^- \rightarrow Ar^+ + 2e^-$; (ii) The positively charged argon or neon ions are accelerated to the cathode, striking it with such force that metal atoms are sputtered from its surface: $M(s) \xrightarrow{Ar^+ (g)} M(g)$; (iii) The vaporized metal atoms are then excited by collisions with electrons or ions: $M(s) \xrightarrow{e^-, Ar^+ (g)} M^*(g)$; (iv) The excited metal atoms return to the ground state and emit the characteristic atomic emission spectrum together with the emission spectrum of the filler gas. Part of the emission spectrum of the Ni-hollow cathode lamp is shown in Figure 21.

The lifetime of a hollow cathode lamp depends, especially, on the consumption of the filler gas and the purity of the cathode material. Filler gas will be absorbed and adsorbed on metal deposited on the tube walls. The decrease of the filler gas depends on the applied current and voltage. The cathode must be made from the purest metal available in order to get the pure emission spectrum. Pure metals are manufactured by electrolytic deposition and during the process hydrogen may be absorbed into the metal. Hydrogen is especially difficult to distinguish from the group 8, 9, and 10 transition metals. Hydrogen in the cathode material will diminish the emission intensity. It also has intensive, continuum emission in the UV region. To get rid of these difficulties lamps are kept in vacuum and at high temperature during their manufacture.

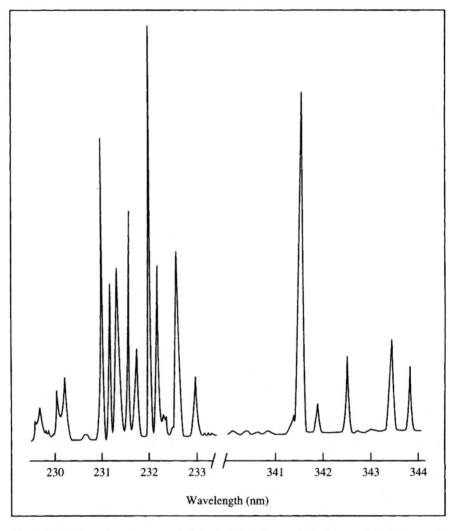

Figure 21   *The emission spectrum of the nickel hollow cathode lamp at 230–233 nm and*
340–344 nm
(Perkin Elmer Corp.)

In modern hollow cathode lamps the electrodes are shielded by
ceramic materials and mica plates in order to avoid energy losses. The
discharge is limited to occur inside the cathode opening. By using
shielding materials, the intensity of the resonance lines increases with
respect to the intensities of the lines of ions and filler gas and the
broadening of the lines decreases.

In principle, in AA analysis an individual radiation source is needed
for every element to be determined. Multi-element hollow cathode
lamps have been designed partly on the grounds of economy and partly

for their convenience. However, it is difficult to prove their economy, because these lamps usually cost more than the single-element lamps, and when the lamp fails, all the elements are lost. On the other hand, the convenience of a multi-element lamp is evident, particularly, in working with a spectrometer which has no lamp 'turret'. In early multi-element lamps cathodes were made of rings of various metals and pressed together. These rings were arranged in order of their volatilities. Nowadays, powders of various metals are mixed in determined ratios, pressed, and sintered. The concentrations of the elements in the cathode alloy must be proportional to their rates of sputtering under the working conditions of the lamp. Otherwise, the different sputtering rates of the elements cause them to lose their intensity and be lost in turn. The elements must be chosen so that the monochromator can separate their resonance lines from each other. Further, the intensities of the resonance lines should be as close as possible to those of each corresponding single-element lamp.

2.2.1.2 Electrodeless Discharge Lamps (EDLs). Among the line-like radiation sources, electrodeless discharge lamps exhibit the highest radiation intensity and the narrowest emission linewidths. They are used both in atomic absorption and fluorescence spectrometry. The biggest advantage of the EDL is the high intensity of the radiation which can be several orders of magnitude higher than hollow cathode lamps, especially, for easily vaporizable elements. The signal to noise ratio is generally improved which gives higher precision and improved detection limits. EDLs are of great advantage, especially, working in the UV region.

The EDLs consist of sealed quartz tubes, 3–8 cm in length and about 0.5–1 cm in diameter. The tube is filled with a few milligrams of the pure analyte metal or its volatile salt (halide) and an inert gas (argon) at a pressure of a few hundred pascals. The lamp is mounted within the coil of a high frequency generator (Figure 22).

The excitation energy is produced by a microwave- (> 100 MHz) or radio-frequency (100 kHz–100 MHz) electromagnetic field. In general, the intensity of the microwave lamps is better than that of radiowave lamps, but the radiowave lamps are more stable and they do not need a thermostat.

2.2.1.3 Microwave EDLs. In the microwave EDLs the pressure of the inert gas must be about 0.13 kPa when the temperature is 480 to 680 K. The material placed in the tube is pure metal or its chloride or iodide. The addition of 1 to 2 mg of mercury or saturation of the tube by mercury vapour prevents the adsorption of the metal, and hence improves the radiation properties of the lamp and adds to its lifetime.

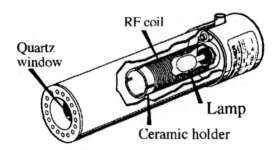

Figure 22   *Construction of an electrodeless discharge lamp*
(Perkin Elmer Corp.)

The intensity of the radiation is controlled by the temperature of the lamp, because the vaporization of metal is mainly thermally induced. Therefore, the microwave EDL must be temperature regulated by a thermostat in order to make the intensity stable enough for analytical use.

These lamps are available for about 50 elements, but the best of them are prepared for arsenic, antimony, bismuth, selenium, and tellurium. On the other hand, the hollow cathode lamps of these elements have the reputation of being the worst lamps which means the hollow cathode lamps and EDLs complement each other.

2.2.1.4   Radiofrequency EDLs.   The intensity of these lamps may be lower than that of microwave lamps, but they give better short and long term stability without the need for a thermostat. The radio frequency commonly employed is 27.12 MHz. Radiofrequency EDLs are available for about 15 elements and they are particularly suitable for routine applications of the volatile elements arsenic, selenium, antimony, tellurium, and phosphorus.

2.2.2   Monochromator

In the AA technique the principal function of the monochromator is to isolate the measured line (resonance radiation) from the other emission lines of the cathode material and lines of the filler gas. Atomic absorption is a very selective method when an element specific radiation source is used. It is practically free of spectral interferences caused by overlapping atomic lines of other elements. The unmodulated molecular emission and other background continua originated in the flame or graphite furnace are eliminated by means of selective amplification. The monochromator consists of two slits (an entrance slit and an exit slit) and a dispersing component (a prism or a grating). The quality of an instrument in other spectrometric techniques, such as in UV-visible spectrometry or plasma-AES, depends mainly on the resolution of the monochromator or on its spectral bandpass. The resolution or resolving power *(R)* of a monochromator is its ability to distinguish adjacent absorption bands or two very close spectral lines as separate entities. Resolution is dependent on the size and dispersing characteristics of the prism or grating, the optical system of the spectrometer, and the slit width of the monochromator. The resolution of any dispersing device is:

$$R = \frac{\lambda}{d\lambda} = w \cdot \frac{d\Theta}{d\lambda} \tag{38}$$

where w is the effective aperture width and $d\Theta/d\lambda$ angular dispersion. The spectral bandpass is the range of radiation that passes through the exit slit.

However, in AAS the resolving power of the monochromator is not so important. Generally, in the AA technique the monochromator should be capable of separating two lines 0.2 nm apart when operating at minimum effective slit width. A monochromator of high resolution is normally not required in routine analysis, like the determination of alkali metals, calcium or magnesium, in biological samples. In principle, no advantages are gained for these kinds of determinations if smaller slit widths than required to separate the analyte line from the other lines of the radiation source emloyed. AAS does not lose its selectivity or specificity if larger slit widths are used, except in the case of complex samples. In analysis of many metallurgical or geological samples, better resolution is required, for instance, for analysing samples containing nickel or chromium. Generally, with decreasing spectral bandpass, the sensitivity increases and the linearity improves. At the optimum slit width (the spectral bandpass at which only the analyte line reaches the detector), a further decrease of the slit width does not bring any advantages.

In contrast to AAS, in plasma-AES the resolving power of the mono-chromator is very important, since emission lines very close to each other must be separated (Figure 135).

The geometric slit width is associated with the effective mechanical widths (in mm or μm) at the entrance and exit slits for a given spectral bandpass. The entrance and exit slits, thus, control the portion of the radiation from the source that enters the monochromator and falls on the detector. By use of a wide entrance slit, large amounts of radiation energy reach the detector. In this case, the noise is small compared to the signal, and lower amplification can be employed. When the noise is low, the signal is stable and precise and low detection limits can be measured. The entrance and exit slits should have very similar mechanical dimensions.

The reciprocal linear dispersion *(dλ/dx)* is the function of the geometric slit width *(S)* and spectral bandpass ($\Delta\lambda_m$) of the monochromator:

$$d\lambda/dx = \Delta\lambda_m /S \qquad (39)$$

The reciprocal dispersion is usually expressed in nm mm^{-1}. For example, if the desired spectral bandpass is 0.2 nm and the reciprocal linear dispersion is 2 nm mm^{-1}, a geometric slit width of 0.1 mm must be obtained. The spectral bandpass of the monochromator in the best quality AA instruments is 0.01 nm.

2.2.2.1 Prism and Grating.   A prism or a diffraction grating is used to disperse the radiation into individual wavelengths. The dispersion of a prism *(dn/dλ)* depends on the refractive index of the prism material *(n)* and the wavelength of the line studied *(λ)*. The dispersion of a prism is high in the UV region, but decreases rapidly with increasing wavelength. Correspondingly, the reciprocal linear dispersion increases with increasing wavelength. The resolution *(R)* of a prism increases with the size of the prism and the rate at which the refractive index of the prism material changes as a function of the wavelength:

$$R = t \; dn/d\lambda \qquad (40)$$

where *t* is the base length of the prism.

Thus, prisms are quite useful in AAS and plasma-AES as the majority of resonance lines lie in the UV region. However, glass prisms cannot be used in these instruments because they do not pass ultraviolet radiation. For example, borosilicate glass is transparent from 310 to 2500 nm, while quartz is transparent from 170 to 2500 nm.

The reciprocal linear dispersion for a grating is nearly constant over the entire wavelength region and it is dependent on the number of grooves per unit width, spectral order, and the focal length of the collimator. The resolution of a grating is a function of spectral order *(m)* and the total number of grooves *(N)*:

$$R = mN \qquad (41)$$

The effective aperture width of a grating is the width of an individual groove *(d)* multiplied by the total number of grooves *(N)* and by cos *r* (*r* is the angle of reflection):

$$w = dN\cos r \qquad (42)$$

For example, a grating ruled with 2000 grooves nm⁻¹ and 50 mm in width has a resolution in the first order of 100 000. At the sodium wavelength of 589 nm, the smallest wavelength interval resolved will be $\Delta\lambda = 589/100\ 000$ nm $= 0.006$ nm.

AAS measurements are almost exclusively carried out in the first order, and a good grating has 2000 to 3000 grooves mm⁻¹. The reciprocal linear dispersion of 1 nm mm⁻¹ or less can easily be obtained with the usual focal lengths employed. Comparable reciprocal linear dispersions are achieved with a conventional prism monochromator only in the far UV region, while in the near UV and visible regions grating monochromators are superior.

Ruled gratings are mainly used in AA instruments. Holographic gratings have less irregularities and therefore lower stray radiation, but this is not so important in AAS as it is, for example, in plasma-AES. From the ruled master gratings (often aluminium) copies can be prepared, which are less expensive.

The various wavelengths are reflected from the grating surface according to the equation:

$$d(\sin i - \sin r) = m\lambda \qquad (43)$$

where d is the distance between the grooves, i is the angle of incidence, and r the angle of reflection (Figure 23). Thus, each line will show up in several reflections. The blaze wavelength, λ_β, is the wavelength for which the angle of reflectance from the groove face and the angle of diffraction from the grating are identical. This value is usually specified as first order wavelength. For instance, the line at 700 nm will appear as the second order line at 350 nm and as the third order line at 233.3 nm. By moving to even higher spectral orders, reflections of different wavelengths start to overlap each other. The dispersion increases with the increasing spectral order (Equation 41). However, this is not a problem, when only first order lines are employed.

2.2.3 Detectors

Photomultiplier tubes (PMTs) are used nowadays almost exclusively in AAS. They are also very common in sequential plasma-AES instruments. A photomultiplier contains a photo-emissive cathode and several anodes (dynodes) in a vacuum. The cathode is coated with an easily ionized material such as alloys of alkali metals with antimony, bismuth, and/or silver.

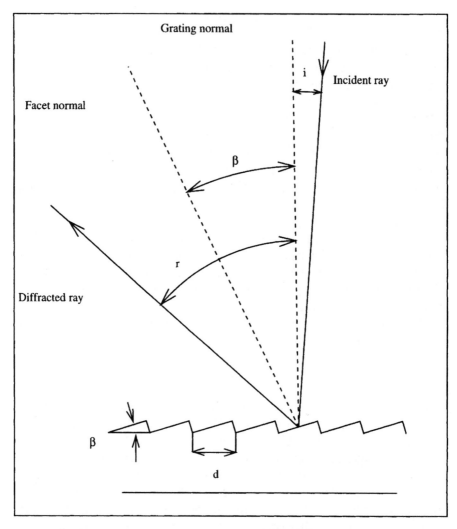

Figure 23   *The reflection of a radiation beam at the grating*

The spectral sensitivity of a PMT depends primarily on the coating material of the photo-emissive cathode (Figure 24). Practically, the lowest measurable wavelength in AAS is 193.7 nm (As) and the highest one 852.1 nm (Cs). In plasma-AES the sensitive lines for sulphur are slightly above 180 nm.

A photon falling on the surface of the cathode causes the emission of an electron, provided the photon is sufficiently energetic to ionize the material. The signal is amplified by the process of secondary emission as shown in Figure 25. The amplification depends on the voltage between the electrodes. The amplification or gain *(g)* increases exponentially

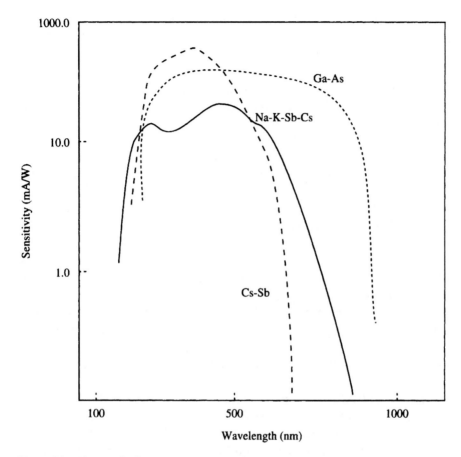

Figure 24 *Photomultiplier sensitivity curves for some cathode materials*

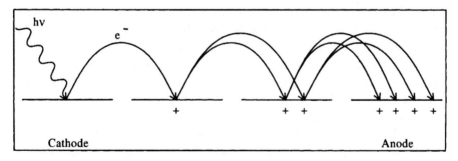

Figure 25 *The signal amplification in a PMT*

with the interanode or interdynode voltage *(U)*:

$$g = kU^{0.7n} \qquad (44)$$

where *n* is the number of anodes. The increase of interanode voltage, however, will increase the dark current and photon noise of the PMT. The current which flows through the PMT under high voltage when no optical radiation is falling onto the cathode is called the dark current. Photon noise is caused by the statistical fluctuation of the photocurrent. Noise from the PMT has a direct influence on the achievable precision and detection limit via the signal-to-noise ratio.

If an electron emitted from the cathode is accelerated to a potential of 100 V, the kinetic energy of the electron emitted will then be about 100 eV (1 eV = 96500 J mol^{-1}).The electron is so energetic that from the first anode it will release 2 to 10 secondary electrons, which are all further accelerated to a second anode which is at another potential 100 V higher. Each secondary electron will again cause further emission of secondary electrons. When a typical PMT contains ten anodes, very high amplifications will be obtained. This can be illustrated by the following example: the photon releases one electron from the cathode and the first anode gives 5 secondary electrons. The second anode gives 5×5 electrons, the third anode $5 \times 5 \times 5$ electrons, *etc.* Thus, with a PMT containing ten anodes, the initial current will be amplified by a factor of 9.8×10^6.

Currently, various types of solid state detectors are coming increasingly popular in the measurement of radiation power, especially when spatially resolved spectrum is utilized in multichannel instruments. These detector devices include photodiode arrays, vidicon detectors and various types of charge transfer devices (CTDs). Especially CTDs are nowadays very popular when multichannel instruments are used in atomic spectrometric measurements. A very good resolution is achieved in these instruments by using an echelle polychromator and higher spectral orders in measurements. The overlapping higher order lines are separated by a prism, resulting in two dimensional spectrum that is simultaneously registered by a two dimensional solid state detector. Echelle polychromators are frequently used in ICP-AES (Figure 142). However, currently also multichannel GF-AAS instruments are commercially available, capable of measuring up to six elements in one atomization cycle (Figure 45).

The charge transfer devices are divided into charge injection devices (CID) and charge coupled devices (CCD). The basic construction of these devices is similar to Metal Oxide Semiconductors (MOSs). Very pure silicon crystal, coated by an insulating SiO_2 layer, is used as a detector substrate. When the incident radiation enters the silicon substrate, localized electron-hole pairs are formed (silicon-silicon bonds are broken if the energy of incident photon is high enough). An appropriate voltage is applied to the conductive overlay (gate), resulting in formation of potential wells within the silicon substrate (Figure 26). Thus a current is generated that is proportional to the photons hitting the detector. Charge transfer devices are composed of individual elements (pixels) that are arranged in two-dimensional order. The number of gates per pixel may vary. The width of individual pixel is typically around 10 μm and the height around 50 μm; the total amount of pixels might be above one million.

There is a principal difference how the information stored is read from CCD and CID detectors. In the former, the charge stored is read sequentially, row by row, and during the read the charge is destroyed. With a CID detector, the accumulated charge is read randomly and it is not destroyed. In addition, it is also possible to read the accumulated charge during signal gathering. CID detectors have higher noise levels than CCD and they have to be cooled to very low temperature (using liquid nitrogen) to obtain acceptable noise levels. CCD detectors are Peltier cooled to −30 . . . −40 °C. They have very low noise levels and low signal intensities can be measured. In fact, the quantum efficiencies of CCDs at different wavelengths are often better than those of PMTs.

The recent advances of CTDs include a segmented array CCD detectors. These devices contain over 200 subarrays, each consisting of few tenths of pixels. These subarrays are positioned to collect spectral information from important areas of two-dimensional emission spectrum obtained by echelle polychromator, instead of reading huge amount pixels in ordinary CCD detector. Each subarray can be read individually and simultaneous background correction can be made. Simultaneous information from different spectral lines of particular element is obtained thus allowing versatile utilizing of collected spectral data

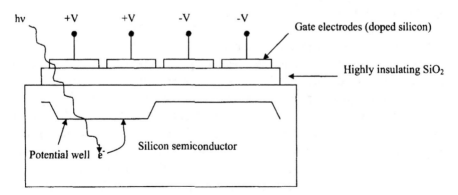

Figure 26 *An example of individual pixel of a CTD having four gates*

(detection of spectral interferences, use of internal standardization, etc.).

2.3 SAMPLE INTRODUCTION IN ATOMIC SPECTROMETRY

Samples to be analysed by different atomic spectrometric techniques and ICP-MS can be in the form of liquids, gases or solids. Samples in liquid form are preferred since they are homogeneous even in a very small scale and also calibration of the measurement is most often straightforward. Generally the analyses are carried out using water as a solvent, but in some cases also organic solvents or mixtures of water and organic solvents are used. By far most of the liquid samples are analysed by introducing them into the measuring device with solution nebulisation. However, liquids may be introduced also as discrete sample droplets. This is the case in graphite furnace atomic absorption spectrometry (and more generally, in electrothermal sample vaporization). When samples in liquid form are used, analytes in solid samples must be first released using a suitable and reliable sample preparation method (Chapter 8).

Sample introduction efficiency of solutions can be considerably improved if an analyte element is first converted to gaseous form before sample introduction. Hydride generation method (for Ge, Sn, Pb, As, Sb, Bi, Se and Te) and cold vapour method (for Hg) are the most universal methods used. Figure 27 describes some widespread ways how liquid samples may be introduced into the atomizer (the scheme shown is by means no complete). Nowadays flow injection (FI) technique (Section 2.3.3) is also often used in connection with different atomic spectrometric techniques. FI provides a very powerful technique for an automated sample manipulation, such as analyte preconcentration, matrix removal, dilution, etc. With the hyphenated techniques speciation analysis is possible (e.g. HPLC-ICP-MS).

The direct introduction of solid samples into the measurement device is a more complicated task than the introduction of solutions. Special, often custom-built, devices are used. The direct introduction of solid samples usually results in poorer precision, and reliable calibration is a major problem. The sample inhomogeneity may also be a problem. However, the dissolution of a solid sample is avoided. The most universal (and commercial) techniques adapted for solid sampling are electrothermal vaporization with a graphite furnace and laser ablation.

It is worth noting that introduction of solid samples as a form of slurries is coming nowadays increasingly popular. In this technique the solid sample is pulverized and nebulized as liquid suspension into

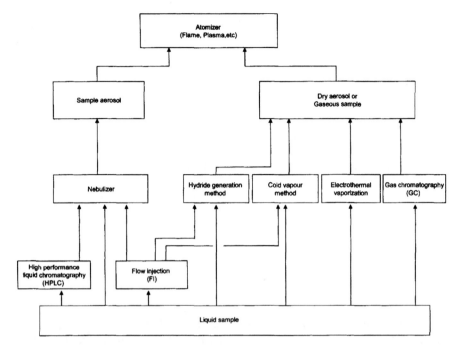

Figure 27 *Introduction of liquid samples into the atomization device for atomic absorption or emission measurement*

atomizer. However, the nebulizers employed (such as V-groove nebulizer) must not be prone to clogging.

2.3.1 Solution Nebulization

In this technique the sample solution is converted to a fine mist using a nebulizer (primary aerosol). After that the large droplets are removed in a spray chamber and solution droplets with size around 10 μm (tertiary aerosol) enter the atomizer. Pneumatic nebulisers are most often used in atomic spectrometry and in ICP-MS. In pneumatic nebulizers the sample aerosol is generated with the aid of a high velocity gas stream. Various types of nebulizers are employed, depending on the sample type (high amount of dissolved solids, high viscosity, etc.) and the measurement technique used (flame AAS, ICP-AES, etc.). The properties of different types of nebulizers, as well as spray chambers used, are discussed more closely under the individual measurement techniques.

Pneumatic nebulizers employing parallel sample and gas flows (concentric nebulizers) are very widely employed in flame AAS, ICP-AES and ICP-MS. Figure 28 shows the principle of the concentric nebulizer that is commonly used in plasma spectrometry. High-velocity nebulizer

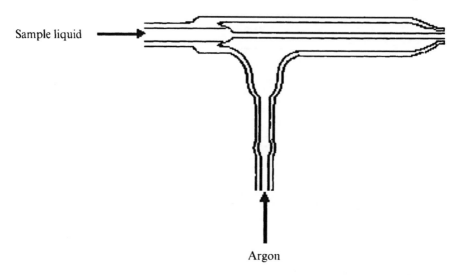

Sample liquid →

Argon

Figure 28 *Concentric nebulizer (Meinhard nebulizer) popular in ICP-AES technique*

gas is introduced through the outer capillary. The pressure drop at the sample capillary draws the sample through the capillary. In the cross-flow nebulizer a liquid-carrying capillary is set at a right angle to the tube carrying the gas stream.

In principle, both nebulizers are self-feeding, and a pressure difference created across the sample capillary draws the sample solution through the capillary according to Poisseuille's equation

$$Q = \pi r^4 p / 8 \eta d \tag{45}$$

where Q is the flow rate of the liquid, r is the radius of the capillary, p is the pressure differential, η is the viscosity of the liquid, and d is the length of the capillary. The velocity of the liquid at the wall of the capillary is assumed to be zero.

The mean diameter of the aerosol droplets produced by a concentric nebulizer (primary aerosol) can be evaluated using an empirical model by Nykiyama and Tanasawa:

$$d_s = \frac{585}{v} (\frac{\sigma}{\rho})^{0.5} + 597 \left(\frac{\eta}{(\sigma\rho)^{0.5}} \right)^{0.45} \left(1000 \frac{Q_L}{Q_G} \right)^{1.5} \tag{46}$$

where d_s is the Sauter mean diameter of the aerosol droplets ($D_{3,2}$, volume to surface ratio) in μm, v is the velocity difference of the gas and liquid at the capillary exits (m s^{-1}), σ is the surface tension of the liquid (dyne cm^{-1}), ρ is the density of the liquid, (g cm^{-3}), η = is the viscosity of the liquid (dyne s cm^{-2}) and Q_L, Q_G are the volumetric flow rates of the liquid and gas (cm^3 s^{-1}). It has been found experimentally by laser diffraction techniques that the empirical model shown gives higher values for the Sauter mean diameter as expected, but it is nevertheless useful in evaluation of the effects of various experimental variables on the constitution of the sample aerosol. Thus the properties of the solvent, such as viscosity affect the droplet size and the amount of sample entering the atomiser.

When the physical properties of the sample solutions and calibration standards are different, this results in altered sample introduction efficiency between them and a matrix effect is observed (Figure 37).

2.3.2 Introduction of Gaseous Samples

2.3.2.1 Hydride Generation. Although reproducible and reliable, one drawback of sample nebulization is the poor sample introduction efficiency obtained. Only about 1–5% of the aspirated sample enters the atomizer and most is going to a waste reservoir. Therefore, to obtain a higher sensitivity, alternative sample introduction techniques are frequently used in atomic spectrometry and ICP-MS.

Eight elements including germanium, tin, lead, arsenic, antimony, bismuth, selenium, and tellurium form volatile, covalent hydrides (Table 5). With the hydride generation method it is possible to separate these elements from the main sample matrix before their introduction into the detection device. The sample introduction efficiency is very high; the amount of the analyte entering the atomizer approach is in principle 100%, and hence much lower detection limits are obtained when compared to ordinary solution nebulisation. Hydride generation technique is most often used together with atomic absorption spectrometry, but it can be equally well coupled with atomic fluorescence spectrometry or with emission or mass spectrometry employing inductively coupled plasmas.

Table 5 *Covalent hydrides of the elements of the groups 14, 15, and 16 of the periodic table, and their average bond enthalpies, melting points and boiling points*

Element	Hydride	ΔH° at 298 K/kJ mol^{-1}	m.p./K	b.p./K
Ge	GeH$_4$	289	107.3	184.7
Sn	SnH$_4$	253	123.2	221.4
Pb	PbH$_4$	—	—	260.2
As	AsH$_3$	297	156.3	210.7
Sb	SbH$_3$	257	185.2	254.8
Bi	BiH$_3$	—	—	251.2
Se	H$_2$Se	312	207.5	231.9
Te	H$_2$Te	267	122.2	270.9

Reduction Methods. Early methods used a zinc – hydrochloric acid reduction (Gutzeit method), together with some kind of collecting device for the hydrides formed:

$$Zn(s) + 2HCl \rightarrow ZnCl_2(aq) + 2H^+ \xrightarrow{M^{m+}} MH_n(g) + H_2(excess) \quad (47)$$

M is the analyte and m may be equal to n or not (for example, As^{III} an As^V are both reduced to AsH_3). Hydrides were collected in U-tubes in nitrogen trap or in rubber balloons. Titanium(III) chloride-hydrochloric acid and magnesium-zinc reductants were used to extend the hydride method to bismuth, antimony, and tellurium. For some elements, especially tin, lead and tellurium, the hydride formation reaction is relatively slow and hence the collection vessel is necessary. In addition, arsenic(V) must be reduced to arsenic(III) by tin(II) chloride or potassium iodide before the actual hydride generation when a metal-acid reduction is employed.

Sodium tetrahydroborate is now generally used as the reductant in various hydride generation methods. The reduction may be illustrated by the following reactions:

$$NaBH_4 + 3H_2O + HCl \rightarrow B(OH)_3 +$$
$$NaCl + 8H^+ \xrightarrow{M^{m+}} MH_n + H_2(excess) \quad (48)$$

The hydride is generated by first adding the sample to a HCl solution (0.5–5.0 mol l^{-1}) and then $NaBH_4$ (about 1% solution). The hydride formation by $NaBH_4$ is very fast and the hydride vapour may be flushed immediately or after reaction times of 10 to 100 seconds into a silica tube atom cell at carrier gas flow rates of 20 to 100 ml s^{-1} when atomic absorption spectrometry is used.

Advantages of sodium tetrahydroborate as the reductant are: (i) The reaction time is fast (varies from 10 to 30 seconds); (ii) $NaBH_4$ can be employed to produce all the eight volatile metal hydrides; (iii) $NaBH_4$ can be added in solution form.

Sodium tetrahydroborate hydrolyses in aqueous solutions in the following way:

$$NaBH_4 + 2H_2O \rightarrow 4H_2 + NaBO_2 \quad (49)$$

The hydrolysis may be prevented by addition of some potassium or sodium hydroxide. The 1% $NaBH_4$ solution may be stored for about 3 weeks.

It is worth noting that tetrahydroborate ion is a strong reductant also in alkaline solution, e.g.:

$$4SeO_3^{2-} + 3BH_4^- \rightarrow 4Se^{2-} + 3H_2BO_3^- + 3H_2O \qquad (50)$$

Hydrogen selenide is a diprotic acid ($pK_{a1} = 3,9$ and $pK_{a2} = 11$) and thus can be liberated by acidification of the alkaline sample. The reaction in alkaline media is useful when for example metal samples are analyzed. It is well known that the ions of transitions metals interfere seriously in the hydride generation. The interference is avoided when the dissolved sample is made alkaline to first precipitate metal hydroxides before the reduction reaction.

Hydride generation method is usually used in aqueous samples. However, the reduction reaction can be carried out also in organic solvents, such as in N, N'-dimethylformamide (DMF) or chloroform. DMF (together with acid) may be utilized for direct solubilization of organic samples (such as plastics) without sample digestion. Sodium tetrahydroborate reductant used may also be dissolved in DMF. The sample preconcentration techniques utilizing solvent extraction of metal chelates with isobutyl methyl ketone or chloroform may be followed with a direct hydride generation from the organic phase. The organic phase is acidified with glacial acetic acid before hydride generation.

Although sodium tetrahydroborate is almost exclusively used as a reducing reagent, there are also other possibilities for hydride formation. One possibility is to use electrochemical reduction instead of ordinary chemical reduction with sodium tetrahydroborate. The analyte elements are reduced in several steps on a cathode made of element that should have very high hydrogen overpotential. Therefore lead is often employed. Platinum is used as an anode. Catolyte sample solution is pumped over cathode and the released hydride is separation by a permeable membrane. One attractive feature of electrochemical reduction is the lower contamination level obtained when compared to the ordinary chemical reduction by $NaBH_4$.

Hydride Generators. Hydride generation methods involve three or four successive steps depending on the technique used: (i) The hydride is generated by chemical reduction of the sample; (ii) The formed hydride may be transferred directly into the atomizer in a carrier gas stream (purge gas) or; (iii) The hydride (if stable enough) can be collected using a suitable collection device, such as liquid nitrogen trap or the hydride may be trapped in graphite furnace; (iv) The hydride is decomposed in the atomizer to form the atomic vapour and the absorption signal is measured. Alternatively, the hydride generation technique can be coupled with atomic emission or mass spectrometry employing inductively coupled plasmas as excitation or ionization sources.

Different modes of operation can be applied for the hydride generation technique: (i) In the normal -'batch' system, the whole sample is reduced in a hydride generator and the hydride formed transported in a carrier gas stream to an detection device; (ii) In continuous mode operation the sample and other reagents are pumped separately by peristaltic pumps and mixed continually in a junction (reaction coil), where the hydrides are formed. The liquid and gaseous reaction products are transported with a carrier gas stream to a gas-liquid separator, and the gas mixture enters the atomiser. A steady state analyte signal is obtained offering the possibility for signal integration; (iii) In the flow injection (FIA) technique all stages of the hydride generation method take place in a fully automated closed system. The FIA system is discussed in Section 2.3.3. Currently continuous flow systems or flow injection systems are almost exclusively used in hydride generation technique. The detection limits of the hydride generation method can be lowered using suitable collectors and short release times for the analytes.

Prior the introduction into the detection device the hydrides must be separated from the sample solution. In the batch system the gaseous hydrides are separated from the liquid phase in the reaction vessel by purging with the carrier gas and/or stirring. In the continuous flow system a phase separator, *i.e.* a gas-liquid separator (Figure 29) or a membrane separator is used. The gas stream entering the detection device also contains substantial amounts of water vapour that may condensate on the walls of the tubing and deteriorates the functioning of the hydride generation system. Therefore the removal of the moisture is

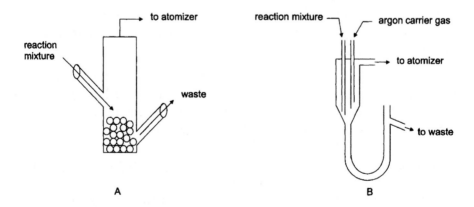

Figure 29 *Two different types of gas-liquid separators used in hydride generation technique. In type A the reaction mixture and carrier gas are mixed beforehand with each other*

usually a necessity. Water can be removed using various drying agents, including concentrated sulphuric acid and drying tubes filled with anhydrous magnesium perchlorate or calcium chloride. However, one has to be aware that the unstable hydrides are not decomposed inside the drying tube. Water can be removed also using a physical approach or a membrane dryer (Figure 30). The latter utilizes a hydrophilic Nafion membrane that adsorbs water. The water is carried away by a stream of a dryer gas.

Interferences in Hydride Generation.   Although very sensitive method, the hydride generation technique is prone to many interferences, and the determinations may result in systematic errors when difficult sample types are analysed (e.g. those containing transition metals). The interference effects observed depend very much on the instrumentation and reaction conditions used, and many published results seem even to be contradictory with each other. In addition to the type of hydride generator used (batch-type, continuous flow or FI system) the chemistry during hydride generation plays an important role.

In principle, the interference effects may occur i) in the liquid phase during hydride generation ii) during the hydride transport to atomizer iii) during the atomization of the hydrides. The first two phases are not dependent of the detection method used and will be discussed below. The interferences during atomisation and detection depend on the device used and will discussed under particular analytical technique.

The interferences effects observed are dependent both the analyte and sample matrix. The concentration and volume of the $NaBH_4$ reductant,

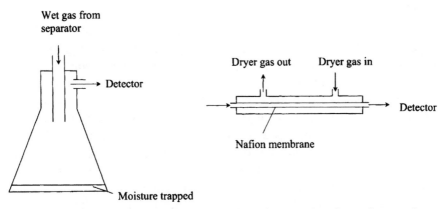

Figure 30   *Moisture removal by means of a physical approach and a Nafion membrane dryer*
(Adapted from W. T. Corns, L. Ebdon, S. J. Hill and P. B. Stockwell, *Analyst*, 1992, **117**, 717)

and the mixing order of the solutions may also affect the magnitude of the interferences effects observed. The chemical form of the analyte (its oxidation state and presence of stable complexes) may slow down or completely hinder the formation of the analyte hydride. Sample properties also play an important role. These include the acids used and their concentrations, and other elements and their compounds present. Especially large concentrations of transition metals cause severe interferences. During the transportation to atomizer hydrides may be decomposed or adsorbed, especially when a drying tube is used between a hydride generator and atomizer.

Kinetic Interferences.   Kinetic interferences are caused by varying rates of formation or liberation of hydride from solution. These interferences occur especially when batch type hydride generation systems are used. Kinetic interferences do not occur when the hydride is collected before being allowed to enter the atomizer. The use of peak area integration instead of the peak height evaluation may eliminate kinetic interferences.

During analysis dense foam is produced when $NaBH_4$ is added to biological samples which have not been fully decomposed. This foam retains a portion of the hydride and the resulting signal is significantly lower and broader than those of acidified standard solutions.

The volume of the sample solution also has an influence on the sensitivity. Hydride is liberated more easily from smaller sample volumes than from larger ones. However, in practice this is not a problem since constant sample volumes are generally employed. In addition, differences in sensitivity between small and large volumes are not very noticeable.

Oxidation State Influences.   The efficiency of hydride formation depends on the oxidation state of the analyte. In the case of germanium, tin, and lead, the oxidation state has no influence on hydride formation.

The sensitivity of As^V is about 70–80% of the sensitivity of As^{III}. For antimony the higher oxidation state, Sb^V, has about 50% sensitivity of the lower state, Sb^{III}. In order to obtain the full sensitivity, a prereduction to the trivalent state is performed by using a mixture of potassium iodide and ascorbic acid or potassium iodide in strongly acidified solution as reductant. The reduction of antimony(v) is almost spontaneous, but arsenic(v) is a little slower, so warming or a longer reaction time is needed. Bismuth exists virtually only in the trivalent oxidation state, so that no prereduction is necessary.

In the case of selenium and tellurium, the difference between the oxidation states IV and VI is very pronounced. The hexavalent state does not give any measurable signal, and prereduction of selenium and tellurium is always required when they exist in the hexavalent state.

Chemical Interferences. Acids generally have little influence on the hydride forming elements except on selenium and tin (Figure 31). Because of strong pH dependence Pb and Sn are determined in buffered solutions. Only hydrofluoric acid causes interferences at relatively low concentrations, so that it should be fumed off before determination. Hydrochloric acid, nitric acid, and sulphuric acid only depress the signal at relatively high concentrations. For most hydride forming elements 0.5 M (1.5%) hydrochloric acid ensures a rapid, homogeneous, and quantitative reaction with NaBH$_4$.

Alkali metals, alkaline earth metals, aluminium, titanium, vanadium, chromium(III), manganese(II), and zinc do not interfere in the determination of hydride forming elements. These metals are frequently present in geological and biological samples, and in sea water and surface water samples. Thus, no major interferences are to be expected when these samples are analysed by the hydride generation technique.

The major chemical interferences are caused by the elements of Groups 8, 9, 10, and 11 of the Periodic Table. A transition metal cation

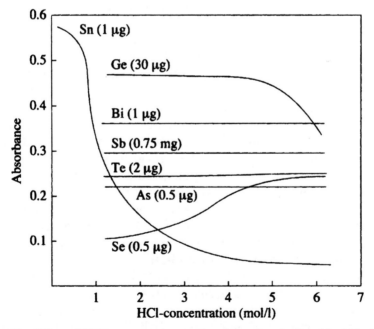

Figure 31 *Effect of HCl concentration on the absorption signals of hydride forming elements.*
(Adapted from: F. J. Fernandez, At. Absorpt. Newsl., 1973, **12**, 93)

present in the hydride generator will be reduced by $NaBH_4$ and precipitated as fine metallic powder (or metal boride). Nickel and metals of the platinum group are hydrogenation catalysts and can absorb a large amount of hydrogen. These metals are also capable of capturing or decomposing the hydride. For example, the addition of 500 mg nickel powder to the sample solution caused total signal suppression. The effect of iron(III) on the determination of antimony is significantly smaller than that of iron(II). The reduction of iron(III) to metallic iron takes place in two steps: first iron(III) reduces to iron(II) which then reduces to metallic iron. In the presence of iron(III) ions, liberation of selenium hydride takes place earlier than the precipitation of iron.

Transition metals may also interfere in ionic form in solutions at high concentrations. It has been concluded that after its formation the hydride will react at the gas/liquid phase boundary with free interfering metal ions in the acidified sample solution giving insoluble arsenides, selenides, tellurides, *etc.*

Interferences caused by transition metals can be relatively easily eliminated or reduced. Interference-free determinations, by the hydride generation technique, have been performed on biological, geological, environmental, and metallurgical samples by increasing the acid concentration of the sample, using an acid mixture, diluting the sample solution, or by using a smaller sample aliquot under otherwise identical conditions.

The Te^{2-} anion is formed by reduction with $NaBH_4$. A number of interfering metal ions form very stable tellurides with this anion. The addition of excess tellurium(IV) to the sample solution allows the interference-free determination of selenium in the presence of Cu, Ni, Pd, Pt, and many other metal ions.

Interferences caused by transition metal ions have been eliminated or reduced by adding various reagents, such as complexing agents, to the sample solution. The addition of these reagents prevents the reduction or precipitation of the interfering metal ion. The interference caused by cadmium, iron, copper, cobalt, and silver on the determination of arsenic can be eliminated with potassium iodide. An addition of potassium iodide reduces the interference of nickel in the determination of arsenic, that of iron in the determination of antimony, and the influence of copper in the determination of bismuth.

EDTA (ethylenediaminetetraacetic acid) forms stable metal chelates with a number of metal ions. Using this reagent as a complexing agent, arsenic, bismuth, and selenium can be determined without any interference in the presence of nickel and cobalt. The cobalt-EDTA chelate is stable in 5 M HCl solution, whereas the corresponding bismuth complex is not. The influence of copper on the determination of arsenic can also be eliminated with EDTA, but not in the determination of selenium. Thiourea has been used to eliminate the influence of copper in the determination of antimony and sodium oxalate to eliminate the influence of copper and nickel in the determination of tin. An addition of thiosemicarbazide and 1,10-phenanthroline reduces the interference of copper, nickel, platinum, and palladium in the determination of arsenic.

Other complexing agents used to eliminate or reduce the influences of metal ions on the determination of hydride forming elements are cyanides, thiocyanates, tartaric acid, citric acid, ammonium fluoride, and 2-mercapto-propionic acid.

Interferences caused by transition elements may also be avoided by separating the hydride forming element from the matrix through co-precipitation with lanthanum hydroxide, iron hydroxide or aluminium hydroxide.

2.3.2.2 Cold Vapour Technique. Another frequently used alternative sample introduction technique in atomic spectrometry is the determination of mercury by the cold vapour technique that is based on the unique properties of this element. Mercury has an appreciable vapour pressure at ambient temperatures (0.16 Pa at 293 K) and the vapour is stable and monoatomic.

The mercury vapour may be entrained in a stream of an inert gas or in air and measured by the atomic absorption of the cold vapour without the need of either flame or flame atomizers. However, with atomic fluorescence spectrometry even lower detection limits are obtained.

The equipments used in cold vapour method are in principle quite the same described earlier for the hydride generation method. Hence batch systems, continuous flow systems and FI-systems are employed. Previously batch-type systems were used, but during the 90's they were gradually replaced with commercial mercury-vapour generators based either on continuous-flow or flow injection principle.

Reduction Methods. Mercury can easily be reduced to metal from its compounds. Mercury is usually reduced to metallic mercury by tin(II) chloride or by sodium tetrahydroborate:

$$Hg^{2+} + Sn^{2+} \rightarrow Hg^0 + Sn^{4+} \tag{51}$$

$$2Hg^{2+} + BH_4^- + 3H_2O \rightarrow 2Hg^0 + B(OH)_3 + 7H^+ \tag{52}$$

Sodium tetrahydroborate is a stronger reducing agent and the reaction rate is faster than with tin(II) chloride. The concentration of used $NaBH_4$ solution is usually between 0.1–0.5% (w/v) compared to 1–10% (w/v) when tin(II) chloride is used. Typically, $SnCl_2$ only reduces inorganic mercury compound whereas $NaBH_4$ also reduces most of the organically bound mercury. Tin(II) chloride is more sensitive to the acid matrix of the samples (acid type and concentration) than $NaBH_4$. A drawback of $NaBH_4$ is its ability to reduce transition metals to elemental states as mentioned earlier. Reduced mercury may be adsorbed by the precipitated metal resulting in systematic errors. Hypophosphoric acid, hydrazine, or Cr(II) also be used as reductants for mercury.

By selective reductions of inorganic mercury(II) compounds and organomercury compounds, it is possible to determine successively the inorganic and organic mercury fraction in the same sample. Inorganic forms of mercury are reduced with Sn(II) chloride and then the organic forms are reduced with Sn(II) chloride – Cd(II) chloride solution.

Enrichment of Mercury.   Due often to very low mercury contents of the samples to be analysed, enrichment procedures are required. Mercury can be extracted with PDDC into MIBK or with dithiazone into chloroform. It can also be electrolysed on the copper or silver cathode. The mercury vapour can be collected in several traps such as various solution traps, gold trap, and active carbon. Adsorption solutions may be nitric acid, bromine water-nitric acid, or potassium permanganate solutions.

Interferences.   Chemical interferences seldom occur in the cold vapour technique. Most acids and very many cations do not interfere even at high concentrations. Interference is caused by Ag, As, Bi, Cu, I, Sb, and Se. The degree of interference is dependent on the reductant used. Only silver interferes equally with tin(II) chloride and sodium tetrahydroborate, while copper, arsenic, and bismuth cause only minor interference with sodium tetrahydroborate. The complete suppression of the mercury signal in the presence of 1 g l^{-1} of Ag, Pd, Pt, Rh, or Ru when using sodium tetrahydroborate as reductant is attributed to reduction of these metals and subsequent amalgamation of the mercury. The same effect can be expected for Cu and Ag. The interference of silver can be avoided by adding bromide. Silver will be precipitated as silver bromide, whereas mercury remains in the solution as the $HgCl_4^{2-}$ complex. The mercury can then be determined free of interference in the filtrate.

Hydroxylamine hydrochloride, which is used to reduce excess permanganate, can have a substantial influence on mercury at higher concentrations.

Complexing agents which form stable complexes with mercury may cause interference. The kinetic interferences can be eliminated if mercury is collected by amalgamation on a noble metal before determination. However, amalgamation does not eliminate chemical interferences.

Sulfur dioxide interferes by absorbing at the mercury line 253.7 nm used for measurements, but SO_2 may be adsorbed into NaOH solution.

The cold vapour technique may be fully automated by using a FIA system, which is discussed in Section 2.3.3.

2.3.2.3 Generation of Other Gaseous Species. The generation of the covalent hydrides mentioned earlier and mercury vapor are by far the most important applications in enhancing the detection capabilities of different atomic spectrometric techniques. In recent years there has been also some progress in the determination of other elements in the presence of $NaBH_4$ reductant. These elements include Ag, Au, Cd, Co, Cu, Mn, Ni, Zn, Os, Pd, Pt, Ru, Rh, Ti and Zn. It is assumed that the elements form hydrides (e.g. CdH_2). However, the compounds formed are very unstable and require very rapid gas-liquid separation and rapid transportation to the atomizer. The generation efficiencies vary, being often well below 100%.

Inspection of the Periodic Table and properties of different elements reveal that also many other volatile compounds that are evaporated at room temperature or at higher temperatures may be utilized in sample introduction. These include volatile chlorides, fluorides, oxides, carbonyls ($Ni(CO)_4$), β-diketonates and other volatile metal chelates.

2.3.3 Flow Injection Analysis

Flow injection analysis (FIA) was first introduced by Ruzicka and Hansen in 1975. FIA is a technique for the manipulation of the sample and reagent streams in instrumental analysis. The purpose of flow injection is to have sample preparation and injection take place automatically in a closed system. The flow injection technique combines the principles of flow and batch type processing and it consists of a set of components which can be used in various combinations.

In atomic spectrometry, FIA has been applied to hydride generation and cold vapour techniques, microsampling for flame atomic absorption, analysis of concentrated solutions, addition of buffers and matrix modifiers, dilution by mixing or dispersion, calibration methods, online separation of the matrix and analyte enrichment, and indirect AAS determinations.

AAS determinations can be fully automated by using a FIA system. For example, by a Perkin Elmer FIAS-200 system up to about 180 determinations can be performed per hour. The FIA system is also economical, since sample and reagent consumption are minimal. Thus also less waste is produced.

2.3.3.1   FIA-FAAS.   Figure 32 shows the principle of the FIA-flame atomic absorption spectrometer. The carrier stream (usually, deionized water or dilute acid) is continuously transported using a peristaltic pump (P1). The sample loop is filled using another peristaltic pump (P2) when the FIA valve is in the 'fill' position. An exact, reproducible sample volume is introduced into the carrier stream and transported into the nebulizer of the spectrometer when the FIA valve is switched to the 'injection' position.

An FIA system is especially useful in the analysis of solutions containing high levels of solids such as saturated salt solutions or dissolved fusion mixtures. The burner slot or nebulizer will not become blocked since the system is continuously and thoroughly rinsed with the carrier stream after each sample measurement. The FIA system requires less than 400 µl of sample solution, which is much less than with continuous aspiration. Thus, FIA-FAAS is the preferred technique when only small sample amounts are available. Low sample consumption is also beneficial for routine analysis, since with fully automated sequential multi-element analysis more determinations can be performed with a given sample volume.

2.3.3.2   FIA-Hydride Generation and FIA-Cold Vapour Techniques. A flow diagram of an FIA system for both hydride generation and cold

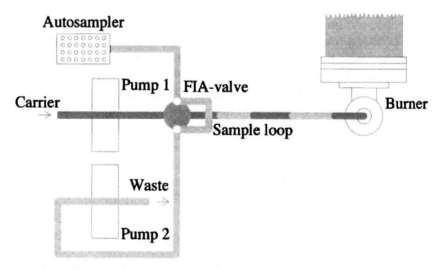

Figure 32   *Principle of the FIA flame atomic absorption*
(Perkin Elmer Corp.)

vapour techniques is shown in Figure 33. The first channel of the peristaltic pump P1 is used for the HC1 carrier stream. When the FIA valve is switched to the 'fill' position, the sample loop is filled with the sample solution by pump P2. Then in the 'injection' position, the sample is introduced into the HC1 carrier stream from the sample loop. The reductant ($NaBH_4$ for hydride generation or $SnCl_2$ for mercury) is transported using the second channel of the pump P1 to the reaction cell where it is mixed with the sample and carrier stream. A spontaneous reaction occurs to form the gaseous hydride of the analyte or to reduce mercury to its elemental form. The reaction mixture is then passed through a gas/liquid separator, where the liquid is separated and removed using the third channel of pump P1. A supplemental argon stream (and hydrogen released during the reaction) transports the gaseous hydrides or mercury vapour to the quartz tube for measurement by AAS. For the determination of hydride forming elements, the tube is electrically heated to decompose the hydrides. In the case of mercury, no heating is required.

2.3.3.3 Amalgamation. The sensitivity for mercury can be significantly improved by using an amalgamation attachment accessory. The mercury vapour is passed through a fine gold/platinum gauze for a specified time. By an amalgamation process, mercury is collected on the

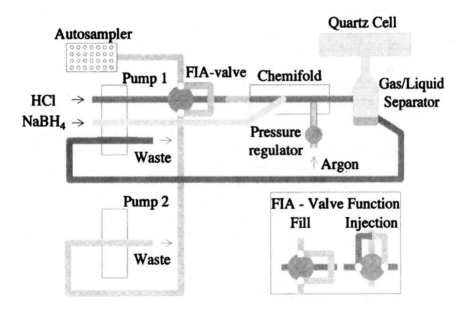

Figure 33 *A FIA system adapted for hydride generation and cold vapour techniques* (Perkin Elmer Corp.)

noble metal gauze. The deposited mercury is then released at once by a rapid heating of the gauze, and the mercury vapour is transported into the quartz tube for measurement. The heating of the gauze may be performed by infrared radiation and cooling below 400 K is carried out using compressed air.

In connection whh the FIA system, the amalgamation attachment allows fully automated on-line concentration of the analyte. This technique improves the detection limit for mercury by an order of magnitude or even more. Gaseous hydrides do not interfere due to the low temperature of the gauze, thus, the analysis with $NaBH_4$ is also interference-free.

2.4 CALIBRATION

In practice, atomic spectrometric techniques and ICP-MS are relative methods and not absolute methods. This means that quantitative results can only be obtained by comparison of sample solutions (or solid sample) with reference solutions (reference materials). The choice of suitable calibration methods and references is important in order to eliminate interferences. Long term fluctuations or changes from one sample series to the next one can be minimized by the choice of the most suitable calibration technique.

2.4.1 Calibration Graph Method

In the calibration graph method, sample solutions are directly com- pared with a set of reference (standard) solutions. The calibration graph is established by plotting the instrument response, e.g. absorbance read- ings for a set of standards against the concentration. In FAAS the 'zero' solution has an absorbance of zero, while highest standard solution has an absorbance equivalent to or more than the highest expected sample solution concentration. It should be noted that ´zero´ solution often in practice contains the element to be determined, for example, when element prone to contamination is measured by GF-AAS (e.g. Zn). Hence a non-zero absorbance is obtained. The calibration graph method is an interpolation method, which means that the readings of all the sample solutions must lie between the lowest and highest readings of the standard solutions. For the highest accuracy, standards are measured before and after the sample solutions. For a large set of samples, standards are measured at regular intervals. The 'zero' solu- tion is usually measured between the standards and samples, to ensure the stability of the base line.

Modern spectrometers determine the calibration function automati- cally and preset the analytical results in desired units (e.g. sample weight and dilution factors are taken into account). In absorption measurements linearity is observed only at a limited concentration range due to unabsorbed light in the system. In emission measurements (such as in ICP-AES) the instrument response is linear over a wide concentration range. However, at higher concentrations nonlinearity is observed due to self-absorption and non-linear response of the detector.

The method of least squares is normally used in the fitting of the calibration data. In its basic form, linear least squares fitting assumes that the random errors in the measured responses are constant at the calibration range used (errors have an equal variance σ^2; homoscedastic data). If this is true, then all calibration points have an equal weight and the errors in the predicted concentration are lowest around the middle part of the calibration graph (Figure 34).

In practice, the variance of the instrument response is not constant (heteroscedastic data). For example, the noise of photomultiplier tube is often proportional to the square root of the signal intensity. Hence, the variances at the different concentration levels should be estimated and a weighted regression line should be used. The weighting factors used are inversely proportional to measured variance at different concentration levels. It can be mentioned that many instrumental techniques (and

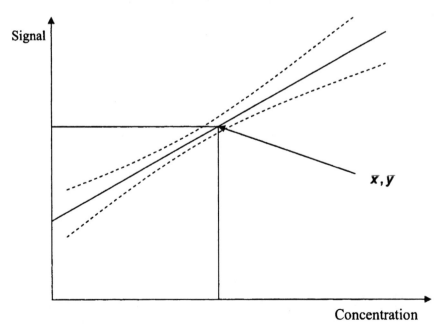

Figure 34 *Confidence limits of a linear calibration function ($y = a + bc$) assuming that the variance of each response is equal*

whole analytical procedures) exhibit a constant relative standard deviation at higher concentrations that simplifies the estimation of the uncertainty associated with the analytical result (Figure 35). A more comprehensive treatment of the calibration data is given, for example, in the textbook mentioned in the "Future reading".

In addition to linear calibration functions, higher order polynomials are easily fitted by the instrument software. The use of microcomputers is advantageous when non-linear calibration functions are encountered. For example, a technique of segmented parabolic functions is used by the Unicam AA instruments (Figure 36). In this technique the blank is measured first, followed by the standards, starting from the lowest concentration. The straight line gradient between the blank and first standard is compared with that between the first and second standards. If the two gradients are within 40% of each other towards the concentration axis, or within 10% of the absorbance axis, the calculation will proceed. Then the mean value of the two gradients is calculated and used to define explicitly a segment of a parabola between the calibration points 1 and 2. When more than two standards are used, the process is repeated between the standard points 2 and 3. The complete calibration function is, thus, built up segment by segment. Excess curvature checks are made at each calibration point. If the preset curvature limits are not met at any part of the graph, an excess curvature error message will be generated.

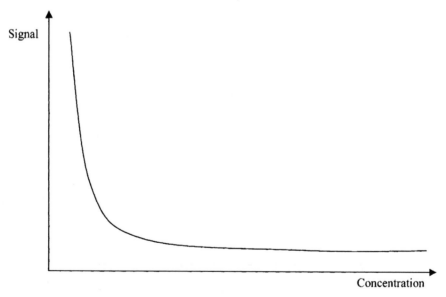

Figure 35   *Typical plot of relative standard deviation (RSD) of the measured instrument response*

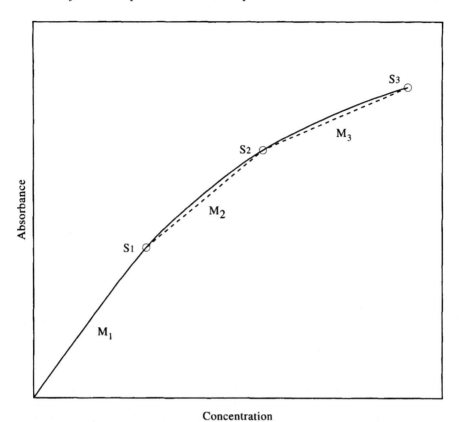

Figure 36 *Segment curve fitting*
(Unicam Analytical Systems Ltd.)

In the analysis of real samples matrix effects (multiplicative errors) are often encountered. The sample matrix may affect the sensitivity of the measurement, *i.e.* the slope of the calibration graph obtained in sample matrix is different to that of synthetic calibration solutions (or calibration samples) (Figure 37). There are numerous reasons for the matrix effects observed, some of them being quite straightforward, such as viscosity difference between samples and standards that affect the sample introduction efficiency. The origin of matrix effects is often dependent on the measurement technique used. For example, the sample matrix may affect the degree of atomization of an analyte in flame atomic absorption spectrometry. In ICP-AES technique the excitation efficiency of an ICP source might be deteriorated by the sample matrix. This may lead to reduced sensitivity, especially for the emission lines that have high excitation and ionization potentials.

Matrix interference (additive error) is encountered when a sample matrix itself generates a signal that is higher or lower to that of pure

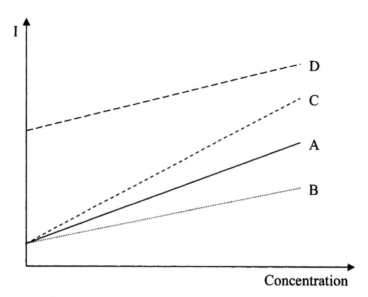

Figure 37   *Examples of how sample matrix may affect the analyte response. Response of pure calibration standards (A); signal depression (B) and enhancement (C) caused by the sample matrix; positive systematic error together with signal suppression (D)*

standard containing no analyte. Hence also the intercept of the calibration graph is changed. Spectral overlaps are a very common glass of matrix interferences that are often encountered in ICP-AES technique.

Matrix effects may be corrected by the use of matrix matched calibration standards or by the use of standard addition method. The effects may also be overcome by diluting the samples, if possible. Matrix interferences often cause serious systematic errors and they might be very difficult to detect. Spiking experiments are often carried out when unknown samples are analysed. However, a complete recovery of an added analyte may be obtained although serious matrix interference exists. Standard reference materials similar to the samples analysed should be used in method development and in routine analysis. Analytical results obtained with a complete different analytical technique are also of a great value. It should be noted that matrix effects and interferences are often dependent on the quality of the instrument used.

2.4.2 Bracketing Method

In this method two standards are selected so that their concentrations closely bracket the expected value for the analyte. This means that only a small section of the calibration graph is used. The concentration of the analyte in the sample is calculated according to the equation:

$$C_a = (A_a - A_{s1})(C_{s1} - C_{s2})/(A_{s2} - A_{s1}) + C_{s1} \qquad (53)$$

where C_{s1} and C_{s2} are the concentrations of the standards s1 and s2, respectively, and A_{s1} and A_{s2} are the corresponding absorbance values.

The advantages of this method are improved compensation of instrument time drift (in contrast to measuring many standards at the beginning of analysis) and suitability to non-linear sections of the calibration graph; provided that the concentrations of the standards are close enough to the concentration of the sample.

2.4.3 Standard Addition Method

It is not possible every time to overcome matrix effects by matching standards with the sample, especially when the full composition of the sample is unknown. For complex samples, where the matrix effects cannot be removed, the standard addition method must be employed. However, this method corrects only for effects which modify the slope of the calibration graph, but not matrix interferences (additive errors).

The sample solution is divided into at least three aliquots and to all but one of these, are added increasing amounts of the element to be determined. The samples are made up to the same volume. The solutions are measured and the absorbance readings plotted against the added analyte concentration (Figure 38). The measured absorbances are due to the analyte concentration in the sample solution, plus the background absorption and scatter. The latter must be corrected by one of the background correction methods (this holds equally well with the background emission observed when ICP-AES technique is used). The true calibration graph would be parallel to the one obtained by standard additions, but passing through zero absorbance. The concentration of the analyte in the sample solution may be read off at C_1 (the intercept on the concentration axis) or C_2, as shown in Figure 38.

The method of standard additions is an extrapolation method, and usually its accuracy is not as good as that of an interpolation method.

Microprocessor-assisted spectrometers have the method of standard additions program included.

2.4.4 Internal Standards

When internal standardization is used, the ratio of the signals of the analyte line and an internal standard line is measured. This technique is well known in ICP-AES, where the internal standard is either the major element in the matrix or an added element of known concentration

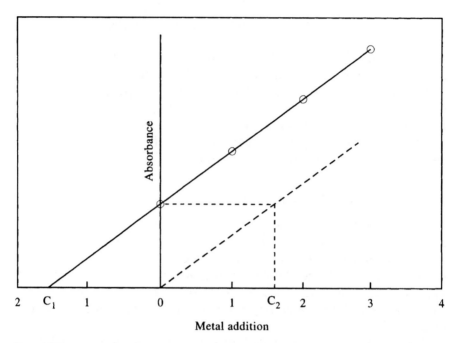

Figure 38 *Standard addition method*

(samples must not contain the element added). To obtain correct results, the element added as internal standard should have exactly the same chemical and physical properties than the analyte element to ensure that they will behave similarly in the analytical system. Small changes in sample physical properties that affect sample introduction efficiency are generally easily corrected by internal standardization. However, if the sample matrix affects the excitation properties of the plasma, the excitation of an analyte line and internal standard line should be affected to same extent. In practice this situation is rarely realized, especially when common internal standard elements, such as scandium or yttrium are used. In AAS, internal standarization may be employed when a double-channel (or multi-channel) instrument is in use and simultaneous measurements at two wavelengths are possible. This technique eliminates small variations in nebulizer and flame performance. Such variations are not corrected by double beam spectrometers.

2.4.5 Calibration Standards

2.4.5.1 Standard Solutions. Relatively concentrated stock solutions are generally prepared for each element to be studied. Working standards are then prepared from these solutions by serial dilution. For the sake of convenience, the concentration of stock solutions is usually

1000 mg l^{-1}. Stocks are prepared either from the pure metals or their simple salts. In routine work, analytical grade reagents are quite sufficient.

Stock solutions are stored in polyethylene bottles, and they are acidified to prevent the hydrolysis of metal ions. Stocks (concentration about 1000 mg l^{-1}) may be stored for several months. Dilute standards (concentration < 1 mg l^{-1}) should not be stored, but always prepared just before use.

Some reagent solutions may be added to standards and samples to overcome interferences, or only to the standards in order to match the samples. These solutions must be of a high degree of purity, especially with respect to the analyte. For example, lanthanum chloride is often used as a spectroscopic buffer in AAS. Ordinary lanthanum salts usually contain traces of calcium and magnesium, the elements which would benefit most by the use of lanthanum in the AAS analysis. However, special AAS grade lanthanum chloride and lanthanum oxide, which are very low in calcium and magnesium, are available from reagent manufacturers. Additionally, a number of other reagents such as alkali metal salts (ionization buffers), EDTA (releasing agents), mineral acids and other dissolution reagents, and extraction solvents are also available in 'AAS grade'. Although blank corrections should be made, the accuracy of the method falls off severely as the blank value approaches the concentration being measured.

The concentration of the stock solution cannot always be obtained accurately enough from weighing, and it must then be determined by a suitable analytical method (for example, complexometric titration).

Nowadays, standard solutions may be obtained commercially from reagent suppliers. The traceability of standard concentrations is also assured by the supplier. In addition to the pure aqueous standard solutions, standards made in varying matrices are also available.

2.4.5.2 Non-Aqueous Standards. When determinations are carried out in organic liquids, standards are prepared by dissolving suitable organometallic compounds in the same solvent. In flame AAS techniques, the solvent has a significant effect on the physical and chemical properties of the flame, and thus has an influence on the precision and sensitivity of the analysis. Only a limited number of organic solvents are suitable for spraying into the flames produced by the usual nebulizer/ spray chamber systems. Many solvents such as aromatic compounds and organic halides are incompletely combusted and produce smoky, yellow flames. Other commonly used solvents (methanol, ethanol, acetone, diethyl ether, lower alkanes) tend to vaporize in the nebulizer and cause an erratic response. The most suitable organic solvents are C$_6$

or C_7 aliphatic esters or ketones and C_{10}-alkanes. In ICP-AES, kerosene or xylene are often used as solvents.

2.4.6 Precision and Accuracy

The measured result should be as close as possible to the true value and the accuracy of the method is connected to the closeness of agreement between the true value for the analyte in a sample and the mean obtained by repeating the analytical procedure several times. The accuracy is then obtained from the difference between the true and measured values. The precision in turn relates to the closeness of agreement, at a given confidence level, between the results obtained by repeating the analytical procedure a large number of times under the same conditions. Accuracy and precision are illustrated in Figure 39.

The standard deviation is given by:

$$s = \sqrt{\frac{\sum_i (x_i - \bar{x})^2}{N-1}} \tag{54}$$

where x_i is an individual measurement, $\bar{x}$ is the mean value for all the measurements, and N is the number of measurements. The relative standard deviation s_r (RSD) is obtained from the standard deviation by

$$s_r = s/\bar{x} \tag{55}$$

According to the IUPAC recommendation, the precision is calculated by taking 3 times the standard deviation (σ) for 30 or more measurements. At the 95% confidence level the factor is 2.83, i.e. there is a 95% probability that the difference of two individual results is not greater than 2.83 σ (repeatability or reproducibility limit).

A number of factors contribute to the uncertainty of individual measurements. The sources of error are random, but systematic errors occur in sampling, in sample preparation, and measurement.

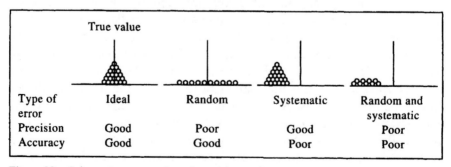

Figure 39 *Schematic presentation of precision and accuracy*

As a consequence of systematic error, the results obtained deviate in the same direction and by about the same amount from the true value. Systematic errors may originate from erroneous procedures at any stage of the analysis and from physical or chemical interferences. To detect systematic errors, interlaboratory tests and/or certified standard reference samples can be used. Mistakes made in sampling cannot be corrected afterwards.

Sources of error in the sample preparation should be recognized and interferences controlled. However, each analysis involves random (statistical) errors, and the whole error is the sum of cumulative errors at each stage of an analytical procedure. With a FAAS, for example, a number of effects contribute to the uncertainty of the final signal displayed on the readout system. In the measurement stage various sources of interference are: fluctuations in radiation source signal, photomultiplier 'shot noise', electronic 'noise', flame fluctuations, nebulization and atomization 'noise', inaccuracies in the read-out system, and interelement interferences.

The standard deviation of the total analytical procedure (σ_T) is given by:

$$\sigma_T = (\sigma_S^2 + \sigma_{SP}^2 + \sigma_M^2)^{1/2} \tag{56}$$

where σ_S, σ_{SP}, and σ_M are the standard deviations originating from sample collection, sample preparation, and measurement, respectively.

When systematic errors do not exist, individual measurements follow a normal distribution. The results give a Gaussian plot where the maximum lies at the mean value ($\bar{x}$) and the shape of the plot is dependent on the standard deviation of the measurement readings.

CHAPTER 3

Atomic Absorption Spectrometry

3.1 INSTRUMENTATION

3.1.1 Basic Features of the Atomic Absorption Spectrometers

The basic components of an atomic absorption spectrometer are shown in Figure 40. The function of the instrument is as follows: (i) The radiation source (a hollow cathode or an EDL) emits a sharp line spectrum characteristic of the analyte element; (ii) The emission beam from the radiation source is modulated; (iii) The modulated signal passes through the atomic vapour where the atoms of the analyte absorb radiation of the line-like radiation source; (iv) The desired spectral line (usually resonance line) is selected by the monochromator; (v) The isolated analyte line falls onto the detector (a photomultiplier) where the light signal is converted into the electric signal; (vi) The modulated signal is amplified by a selective amplifier; (vii) The signal is finally recorded by a readout device (a meter, strip chart recorder, or through data processing to a digital display unit or printer).

3.1.2 The Lambert-Beer Law in Atomic Absorption Spectrometry

The Lambert-Beer law is usually derived in the following way: the intensity of a monochromatic radiation emitted from the radiation source is I_o. Passing through the atomizer with a length of b, the intensity of the light will decrease to the value I. The reduction of the light intensity in the length db is $-dI$ which is directly proportional to the light intensity I, to the length db, and to the number of absorbing atoms N. The number of atoms is in turn directly proportional to the concentration of the metal in the test solution (c):

$$-dI = kIcdb \qquad (57)$$

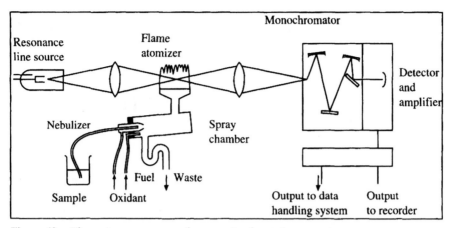

Figure 40   *The main components of an atomic absorption spectrometer*

where k is the relative coefficient. Now the Lambert-Beer law is obtained by integrating over the whole length of the atomizer:

$$-\int_{I_o}^{I} dI/I = kc\int_{o}^{b} db \qquad (58)$$

or

$$\log(I_o/I) = k'cb \qquad (59)$$

The Lambert-Beer law is normally written as:

$$A = abc \qquad (60)$$

where A is the absorbance $[\log(I_o/I)]$ and a is the absorption coefficient. In a series of measurements b is constant and determines, together with the absorption coefficient, the slope of the calibration graph, *i.e.* the sensitivity of the method.

Between the absorbance and transmittance (T) is the following relationship:

$$A = log(I_o/I) = \log(1/T) \qquad (61)$$

The output of the detector is determined by the energy falling on it, and is directly proportional to the transmittance. Transmittance values are then converted to absorbance values by calculation, by a non-linear scale on the instrument meter, or electrically in instruments with linear absorbance readout.

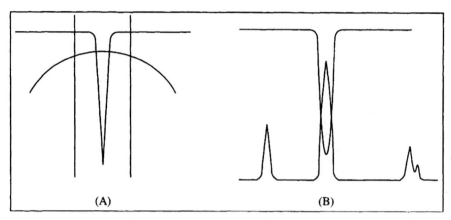

Figure 41   *(A) Absorption line and continuum emission; (B) Absorption line and sharp line emission*

It is important in AA measurements that the emission line width coming from the radiation source is narrower than the absorption line width of the atoms studied. In principle, a high resolution monochromator is not needed to separate the analyte line from the other lines of the spectrum, but in practice, the spectral bandpass of the source should be equal or less than the absorption line width. Otherwise, artificially low absorbance values are obtained leading to reductions in sensitivity. In the AA technique the use of continuum sources (quartz-halogen filament lamps and deuterium and xenon arc lamps) with reasonably priced monochromators is not satisfactory. This is demonstrated in Figure 41 In the case of (A) the emission of radiation is continuous for the whole spectral bandwidth. The energy absorbed by the atoms of the analyte is small in comparison to the whole energy falling onto the detector. In this case much light energy is passing through the monochromator slit and the measurement becomes inaccurate. In the case of (B) by using a line-like radiation source the situation is opposite and the emission and absorption profiles overlap each other. The most important radiation sources in atomic absorption spectrometry are the hollow cathode lamps and electrodeless discharge lamps. Other sources which have been used are lasers, flames, analytical plasmas, and normal continuum sources like deuterium and xenon arc lamps.

Three possible calibration curves observed in atomic absorption spectrometry are shown in Figure 42: (A) This is an ideal plot, where the absorbance obeys the Lambert-Beer law in the whole concentration region; (B) This is a normal curve, which is first linear and then bends against the concentration axis; (C) This is an occasional curve, which is

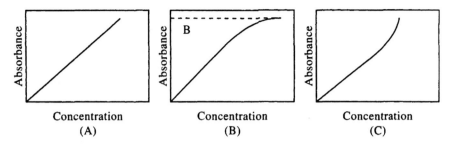

Figure 42 *Types of calibration graph. A: ideal; B: normal; C: complex*

first linear and then shows a curvature away from the concentration axis.

The normal calibration curve is asymptotic to the value of (B), which represents the transmission of unabsorbed light. Unabsorbed light may be due to stray light or non-absorbing lines from the radiation source (cathode material or filler gas) which pass within the spectral bandwidth of the monochromator. The slope of the calibration graph is also dependent on the ratio of the half-widths of the emission line (w_e) and absorption line (w_a): (i) The curve is linear when $w_e/w_a < 1/5$; (ii) The curve is slightly curved when $1/5 < w_e/w_a < 1/1$; (iii) The initial slope starts to decrease when $w_e/w_a > 1/1$. Calibration graphs may have curvatures away from the concentration axis (Figure 42C) when, for example, sodium, potassium, and gold are determined in an air-acetylene flame, or barium and europium in a dinitrogen oxide-acetylene flame. This is due to ionization of the analyte. The ionization decreases with increasing concentration. This is also seen in Figure 56 .

3.1.3 Scale Expansion

Scale expansion is used to measure very small concentrations (signals less than 0.1 absorbance units) with a higher degree of confidence. A scale expansion facility of up to 100 times is often provided in commercial spectrometers. However, factors of more than 20 times are not normally useful, since by a scale expansion noise increases, and longer integration times or the mean of several individual measurements must be used.

3.1.4 Single-beam and Double-beam Instruments

Figure 43 shows the light path in a single alternating current (AC) spectrometer. The radiation which falls onto the detector is

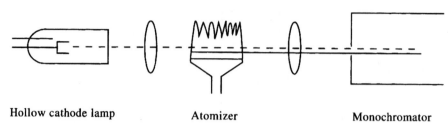

Hollow cathode lamp   Atomizer   Monochromator

Figure 43 *Schematic construction of a single-beam AA instrument. Dotted line represents modulated signal from the radiation source and solid line direct current emission from the atomizer*

proportional to the transmittance. Therefore, it is necessary to run two measurements (sample and blank solutions) in order to get the corresponding absorbance reading. With a single-beam system accurate absorbance measurements are obtained provided that the radiation source is very stable. Although the emission intensity of every radiation source will change as a function of time, the modern hollow cathode lamps and EDLs meet this requirement well. The principle of a double-beam spectrometer is shown in Figure 44. The emission signal from the radiation source is divided by a rotating mirror-chopper into the sample and reference beams. The sample beam is crossing through the atomizer and the reference beam is by-passing it. The two beams are then recombined by a semi-transparent mirror behind the atomizer. The electronics of this system yields the ratio of these beams with the sample beam as the denominator and the reference beam as the numerator.

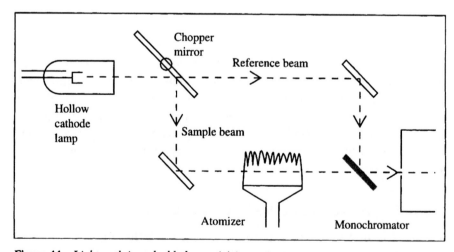

Figure 44 *Light path in a double-beam AA instrument*

Since the sample and reference beams are both generated from the same radiation source, passing through the same monochromator, and received by the same detector and amplified by the same electronics, any variations in the radiation source, sensitivity of the detector, or amplification appears in both beams and cancels out. The stability of this system is thus better than that of a single-beam system. On the other hand, as the reference beam does not cross the atomizer, variations in the atomizing conditions cannot be eliminated. A further disadvantage of double beam optics is the loss of incident energy (about 50%), which is much higher than in single-beam optics.

3.1.5 Multi-channel Instruments

In the dual-channel system two radiation sources are used. The beams from these sources are crossing the same atomizer and then they are handled independently by two monochromators, two detectors, and two electronic units.

These systems are very versatile. Absorption and emission measurements can be run by both channels. A dual-channel system permits the simultaneous determination of two elements. The sample amount and time needed are then only half of that required by a conventional AA system. This is of great use in analyses of small samples. The background correction can be performed by the other channel. Then the difference of the absorbance readings of the two channels is measured. It is also possible to obtain the absorbance ratio of the two channels, which can be used in working with the internal standards.

Currently also multi-channel FAAS and GF-AAS instruments are commercially available (Figure 45). Several line sources are used and the combined beam passes through the atomizer. The analyte elements in the atomizer absorb their own characteristic radiation. The instrument is equipped with an echelle polychromatror and a solid state detector for simultaneous absorption measurement. Up to six elements can be measured simultaneously with graphite furnace atomization. Hence the long analysis time due slow atomization cycle in GF-AAS is considerably improved with simultaneous determination.

Multi-channel atomic absorption spectrometry has some limitations. Any combination of the analyte elements during the analysis is not possible; for example spectral overlaps must be taken into account. Compromise must be made with the analysis program used. In an unknown sample the concentrations of the analytes may vary and the fixed sample volume used may cause problems since the sensitivities

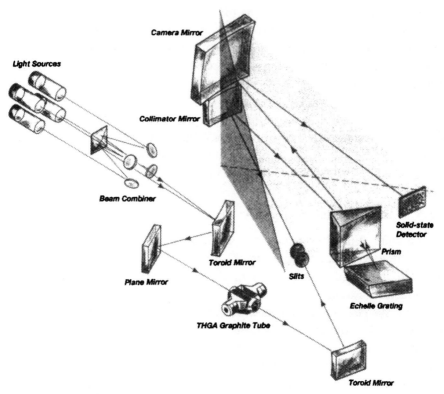

Figure 45   *Simultaneous graphite furnace AA spectrometer*
(Perkin Elmer Corp)

and optimal concentration ranges for different elements are not alike.
To obtain best results elements exhibiting similar thermal behavior
should be measured with the particular furnace program employed,
since optimum ash-atomize plots for individual elements may vary
considerably (Figure 81). Matrix modification is an essential part in
multi-channel GF-AAS determination.

3.1.6   Modulation of the Signal

The optical signal falling upon the detector consists of the resonance
line radiation and emission from the atomizer. The emission originated
from the flame or graphite furnace consists of molecular band emission
and scatter from small particles. Only the resonance signal of the
radiation source is wanted, since other radiation falling on the detector
diminishes the absorbance value that can be recorded. The output of
the radiation source is therefore coded by modulation, and an amplifier

placed after the detector is tuned to the same modulated frequency. The radiation source can be modulated either electrically or mechanically. Electric modulation can be performed by using an AC supply current to the lamp (frequency of 280 or 300 Hz, for instance). In mechanical modulation, the radiation beam is interrupted using a rotating sector (chopper) before the atomizer. This method is used, especially, in the double-beam instruments. The chopper splits the light beam into sample and reference beams at a fixed frequency. A beam of regularly varying intensity is produced which will generate an alternating signal at the detector. An AC amplifier is tuned to amplify the signals only at the same frequency as the beam modulation. Thus, all the noise at other frequencies is rejected, and the signal-to-noise ratio will improve.

3.1.7 Readout Devices

Meters with calibrated scales were used for readout in early AA instruments. Nowadays, digital displays and graphical presentation of signals on the screen of video display units are used. Analytical results can be made as a hard copy on a chart recorder, a printer, or a plotter. A typical hard copy of the analytical results is shown in Figure 46.

3.1.8 Optimization of Operating Parameters

The fundamental function of atomizers in AAS is simply to produce atoms. The basic steps carried out continuously in flame atomization are: (i) Nebulization (conversion of sample into droplets leaving small particles); (ii) Desolvation (removal of solvent); (iii) Atomization (thermal or chemical breakdown of solid particles); (iv) Measurement (interaction with radiation); (v) Condensation of reaction products (residues removed by exhaust flame gases). These same steps are carried out in electrothermal atomization, but in discrete, sequential steps: (i) Injection (injection of the sample into the graphite tube); (ii) Drying (drying the sample into solid particles); (iii) Thermal pretreatment (process of removing volatile matrix constituents to reduce interferences and background); (iv) Atomization (thermal or chemical breakdown of solid particles); (v) Measurement (interaction with radiation); (vi) Tube clean (process of removing refractory constituents of the sample matrix).

A number of instrumental parameters will affect precision and sensitivity in these atomization techniques. In both systems, bandpass (slitwidth) and lamp current must be optimized. In flame AAS, the

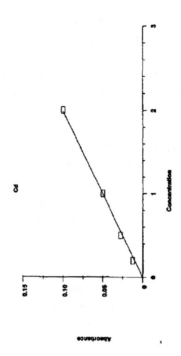

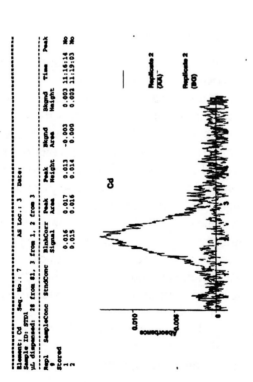

Figure 46 *A typical print out of a graphite furnace instrument*
(Perkin Elmer Corp.)

other operating parameters are burner height, burner alignment, fuel flow, flame type, and impact bead adjustment. The use of background correction and choice of integration time will affect the precision. The sensitivity may be modified by the use of alternative lines and switching out of the impact bead. An appropriate spectroscopic buffer can be used to overcome interferences which reduce sensitivity.

In the electrothermal AAS system each step in the temperature programme of the graphite furnace must be optimized. Each stage requires different temperatures, times, and ramp-rates depending on the analyte and the sample matrix.

3.1.9 Automation

Automation can be used in AAS to make easier and faster, the various steps of analyses. For instance, the following operations can be automated: (i) Sample preparation; (ii) Sample dilution; (iii) Sample introduction; (iv) Multi-element analysis; (v) Data handling.

3.1.9.1 Sample Preparation. The preparation of the final sample solution from a solid sample is often the most time-consuming step of the analytical procedure. This is also the most difficult operation to automate. Weighing, digestion, filtering, and dilution of the sample are procedures which are difficult to perform completely without human intervention.

For example, the automated sample preparation procedure of geological samples could be the following: (1) Sample tubes are placed on a stand; (ii) Samples are weighed with a balance connected to a computer. Sample stands with the tubes are placed one after another on the balance. Each weighing series and sample tube is coded; (iii) An acid mixture is added into the tubes; (iv) Stands are transferred to the programmeable digestion system. This is the slowest step, and it is often convenient to perform during the night time according to a preset temperature program; (v) Samples are measured using an autosampler.

3.1.9.2 Sample Dilution. Commercial equipment is available which enable predetermined volumes of any number of samples to be accurately diluted with another liquid (solvent, buffer, *etc.*). The use of this equipment saves considerable time that would be spent pipetting and making up to volume large sample series.

3.1.9.3 Sample Introduction. The functioning of a conventional autosampler in flame AAS is the following: (i) When the sample tube is in 'analysis' position, sample take-up is initiated; (ii) After a suitable delay, while the absorption signal is stabilized, the sampler gives a signal to the main instrument to carry out the peak integration and data handling steps; (iii) After the measurement, the main instrument signals to the sampler to proceed to the next sample and a similar sequence takes place.

In electrothermal atomization the operation of autosamplers is somewhat different. The non-continuous nature of the sample introduction for GF-AAS requires a wash-through for the sampling capillary between samples in order to avoid contamination. The sample is then taken up and transferred by an autopipette into the atomizer. Then the sampler gives a signal to the furnace controller to start the preset heating program. During the cooling step the controller signals to the autosampler to change to the next sample and carry out the wash-through.

Modern autosamplers are capable of making the calibration, dilutions, the method of standard additions, and matrix modification completely automatic. The appropriate series of blanks, standards, and reagents are loaded and the data handling system programmed. Recalibration is also possible at selected intervals during a long sample run. Using the autosampler, precision can be improved considerably compared to that achieved by hand pipetting.

3.1.9.4   Multi-element Analysis.   The low productivity due to ordinary single-element nature of atomic absorption measurement can be improved by using automation. Modern instruments can automatically switch between different analytical conditions needed for various analytes in a particular sample type. Thus the selection of the correct lamp from those available in the turret and the lamp current is automatically done. Also other instrumental parameters such as slit width, burner height and fuel flow defined by the user or instrument manufacturer can automatically setted. The instrument software will also monitor the quality of the results and recalibration, for example, is carried out if needed. Hence rapid sequential measurement of many elements is possible with little man power, particularly with a rapid flame atomization.

3.1.10   Fault Finding

The goal of AAS measurements is to obtain results which are both accurate and precise (Figure 39). However, a number of problems are encountered with AAS techniques, most of which may be solved by the analyst. There are four main problem areas to be considered: (i) Low sensitivity; (ii) Poor precision; (iii) Drift; (iv) Incorrect results.

3.1.10.1   Flame AAS.   Table 6 lists various reasons for low sensitivity and poor precision in FAAS.

Table 6 *Causes of low sensitivity and poor precision in flame atomic absorption spectrometry*

Cause for low sensitivity	Cause of poor precision
	Flame on
Incorrect flame type or conditions	Partially blocked nebulizer
Incorrect lamp	Partially blocked burner
Incorrect lamp current	Burner mis-aligned
Burner mis-aligned	Dirty spray chamber
Impact bead out of adjustment	Moisture trap in air supply full
Blocked nebulizer	Inadequate drainage of spray chamber
Blocked burner	Acetylene cylinder pressure less than 70 Pa
Incorrect slits	
Incorrect capillary uptake tubing	
Incorrect standard solution	
Particles greater than 10 μm	
Incorrect chemistry (no ionization buffer)	
	Flame off (electronic)
	Incorrect lamp current
	Incorrect slits
	Incorrect lamp alignment
	Incorrect wavelength
	Dirty optics
	Background correction selected at high wavelengths
	Background correction/HCL energy mismatch
	Faulty HCL
	High scale expansion

Incorrect standard reference material, incorrect standard preparation, and incorrect sample preparation can cause incorrect results. In addition, if background correction is not employed when required, it will give rise to erroneous results.

When it is difficult to light the flame, a distorted igniter or incorrect gas pressure may be the causes. If the instrument will not change to the N_2O-acetylene flame, the reason might be that the wrong burner is fitted or there are incorrect gas pressures. If the lamp will not align, the reason might be: incorrect nose cone, lamp has failed, faulty lamp lead or power supply, gain overload, or incorrect display.

3.1.10.2 Electrothermal AAS. Many problems to be considered are common to GF-AAS and FAAS. In GF-AAS it is essential that the furnace parameters are also optimized for a given analysis. The most common causes of low sensitivity are: (i) Sample drying too quickly; (ii) Too high a pretreatment (ashing) temperature; (iii) Too low an atomizer temperature; (iv) Not using temperature control during the atomization stage; (v) The inert internal gas supply is still flowing during atomization; (vi) Sample spreading in the tube. Various types of interferences will also cause reduced sensitivity. Interferences in GF-AAS may be classified as follows: (i) Physical interferences; (ii) Stable compound formation; (iii) Volatile compound formation; (iv) Vapour phase interferences. Interferences in electrothermal atomization are discussed in detail in Section 3.3.8. It is also essential that the correct cuvette (electrographite, pyrolytically coated, totally pyrolytic, or ridge cuvette) is selected for the element/matrix to be analysed.

The most common causes of poor precision include faults in sample delivery system, injection depth, furnace programme, radiation source, background correction, or optics. The volume of solution injected into the furnace must be reproducible. However, imprecise injection is unlikely to occur with modern autosamplers, but is a problem with manual pipetting. Incorrect injection depth is a common cause of poor precision in GF-AAS. As a general rule, the more viscous samples are injected closer to the tube wall. Sample spreading during the drying stage is a common problem with organic solvents. If the furnace is misaligned with respect to the background corrector, problems will be seen due to either attenuation of the light source or from tube wall emission. The signals recorded may be noisy due to attenuation of the light source or the use of incorrect slits. Dirty optics, especially with double beam systems, will reduce light throughput.

There are many reasons for incorrect data, but many times the source of error is traceable to either standards, sample preparation, or contamination.

3.2 FLAME ATOMIZATION

The emission spectrum of the analyte emitted from the radiation source is passed through an atomizer in which a portion of this radiation is absorbed by the free atoms of the analyte. The main function of the atomizer is to produce ground state atoms of the element to be determined from the ions or molecules in the sample. This is the most difficult and critical step in the AA procedure. The sensitivity of the determination is directly proportional to the atomization degree of the analyte element, which is in turn dependent on the effectiveness of the atomizer. The longest and most widely used atomization technique in AAS is spraying of the sample solution into a flame. Electrothermal techniques (especially, graphite furnace AAS), the hydride, and cold vapour methods, as well as various 'semiflame' techniques are all very important atomization methods nowadays, especially for determinations of metals at ultratrace concentrations. All these different atomization techniques do not compete, but complement each other in an ideal manner.

3.2.1 Flames

In the flame atomization technique the sample is sprayed into the flame in the form of an aerosol generated by means of a nebulizer. It is also possible to introduce solid samples directly or as suspensions into the flame.

Flames employed in AAS may be divided into two groups: the combustion flames and diffusion flames. In fuel-oxidant mixtures, the temperature of the flame varies generally from 2000 to 3000 K. Air and dinitrogen oxide (N_2O) are the most widely used oxidants, and acetylene, propane, and hydrogen are the most common fuel gases. In diffusion flames, the fuel is also the carrier gas and it burns on coming into contact with the outer diffusion air. The temperatures of the diffusion flames are lower than those of the combustion flames. The characteristics of some commonly used flames are given in Table 7.

3.2.1.1 Combustion Flames. The atomization is dependent on the temperature and the chemical environment *(i.e.* chemical compounds and radicals existing in the flame) of the flame. The burning temperature of two different gas mixtures can be the same, but their analytical characteristics could be quite different.

Pure oxygen is rarely used as an oxidant because its burning velocity is high and it is difficult to control. However, oxygen may be mixed with argon or helium to overcome these disadvantages.

Most of the gas mixtures summarized in Table 7 can be used in varying ratios which may be stoichiometric, lean, or rich. Lean and rich mixtures contain less or more than the stoichiometric quantity of fuel gas, respectively.

The most widely used flame is the air-acetylene flame. For about 30 elements it offers a suitable environment and a temperature sufficient for the quantitative determination. In a few cases some ionization occurs in the air-acetylene flame, like in the determination of alkali

Table 7 *Characteristics of some common flames*

Flame	Flow rate (l min⁻¹)		Temperature(K)	Burning velocity (cm min⁻¹)
	Fuel gas	*Oxidant*		
Air-propane	0.3–0.45	8	2200	45
Air-acetylene	1.2–2.2	8	2450	160
Air-hydrogen	6	8	2300	320
N_2O-propane	4	10	2900	250
N_2O-acetylene	3.5–4.5	10	3200	285
N_2O-hydrogen	10	10	2900	380

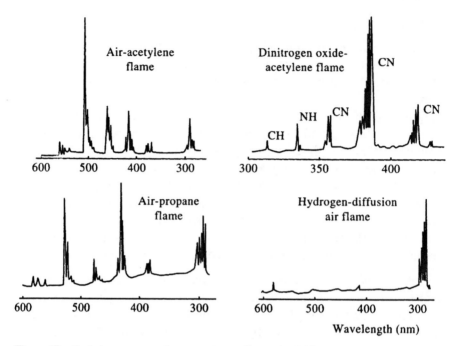

Figure 47   *Emission spectra of some common flames in AAS*
(Adapted from: M. S. Epstein and T. C. Rains, Anal. Chem., 1976, **48**, 528)

metals. The flame is transparent over a wide spectral range and shows only noticeable radiation absorption below 230 nm. Also, the emission of this flame is very low giving ideal determination conditions for many elements (Figure 47). Calcium, magnesium, strontium, chromium, iron, cobalt, nickel, molybdenum, and noble metals are normally determined with the air-acetylene flame. The alkaline-earth metals are determined with a rich flame (reducing) whereas the noble metals (rhodium, iridium, palladium, platinum, gold) are determined with a strongly lean (oxidizing) flame.

The temperature of the air-acetylene flame is nevertheless insufficient to determine a number of elements which form thermally stable oxides. These elements are called refractory oxide elements. The sensitivity of the determination of refractory oxide forming elements is higher in the rich flame than in the stoichiometric flame, although the former one is about 150 K cooler. Elements having dissociation energies for the metal-oxygen bond greater than about 5 eV, give low sensitivity in the air-acetylene flame. Typical refractory oxide elements are aluminium, titanium, zirconium, and tantalum which have the following M-O bond dissociation energies: Al-O 5.98 eV; Ti-O 6.9 eV; Zr-O 7.8 eV; and Ta-O 8.4 eV.

The air-propane flame was commonly used in the early days of AAS for those elements which can be easily atomized (alkali metals, cadmium, copper, lead, silver, and zinc). The air-propane flame is not widely used except in laboratories where the use of acetylene is forbidden. The dinitrogen oxide-propane flame is more advantageous for use in these cases because it possesses less chemical interferences and it allows the determination of some refractory oxide elements.

The dinitrogen oxide-acetylene flame can be used for those elements which cannot be determined successfully with an air-acetylene flame. The temperature of the N_2O-acetylene flame is higher than that of the air-acetylene flame. In addition, the air-acetylene flame is safer because of its smaller burning velocity (Table 7). A slightly rich N_2O-acetylene flame consists of about 2 to 4 mm high blue-white primary reaction zone, above that about 5 to 50 mm high red reduction zone, and on the top a blue-violet secondary reaction zone where the fuel gas oxidizes. The dissociation of the sample takes place in the red reduction zone.

Sensitivity is diminished due to many elements ionizing to some degree in the hot N_2O-acetylene flame. Ionization problems can be reduced by adding some other easily ionizing element in excess.

The emission from the N_2O-acetylene flame is strong. CN-, CH-, and NH-emission bands exist in a wide spectral range (Figure 47). The coincidence of some of these bands with the resonance radiation will affect emission noise, which in turn decreases precision.

Hydrocarbon flames contain CH_3, CH_2O, CH_2, CO, H_2O, OH, and H molecules and radicals. If N_2O is used as the oxidant, CN, NH, and NO radicals are also formed.

The N_2O-acetylene flame demands a shorter burner slot than the air-acetylene flame. Generally, the size of the burner slot must be such that the flow velocity of the gases becomes greater than the burning velocity of the gases. For safety reasons it is important to control the flow rates of the gases and to use the proper burner for each type of gas mixture.

The critical gas flow rate of a laminar burner (the minimum rate at which no explosion occurs) is only about 2 to 3 times greater than the burning velocity of gas mixture. The flame is ignited usually with a small excess of the fuel gas, then the burning velocity is small and the gas flow rate fast. The N_2O-acetylene flame is ignited and put off via air-acetylene.

In Figure 47 the emission spectra of some commonly used flames are shown, and in Figure 4 elements which are recommended to be determined with an air-acetylene or a N_2O-acetylene flame are summarized.

3.2.1.2 Diffusion Flame. The strong radiation absorption of low wavelength hydrocarbon flames is why hydrogen is used as the fuel gas in many applications. Hydrogen flames are very transparent over the whole spectral range (Figure 47). They are also safe, because gases are not mixed before burnt. In these flames, the fuel gas also serves as the carrier and it burns on coming into contact with the outer diffusion air at the opening of the burner slot.

For the hydrogen-diffusion air flame the same burners are used as for the air- or dinitrogen oxide-acetylene flames. An inert gas (argon or nitrogen) is mixed with hydrogen to improve the analytical characteristics of the flame. Then the sample is sprayed into the spray chamber with the help of the inert gas. Diffusion flames are employed for determination of easily atomizing elements, such as arsenic or selenium.

3.2.2 Nebulizer-burner Systems

The main function of the nebulizer-burner systems is the conversion of the sample solution first into the aerosol and then into the atomic vapour, after which the absorption measurement is carried out. These systems are the most important part of the AA instrument. Their correct function and efficiency guarantees the success of the analysis.

Nebulization includes the conversion of the sample solution into a mist or aerosol. In pre-mix spray chambers, mist droplets of the correct size distribution are selected and introduced, with the flame gases, into the burner in which the atomization takes place. The technique most often employed to produce atomic vapour is the indirect nebulization system described above. The method in which the nebulizer and burner are combined is called a direct nebulization. In this method the sample mist is formed and mixed with the fuel gas at the opening of the burner. Burners employed in the indirect and direct nebulization systems are called pre-mix burners and direct injection burners, respectively.

A typical pre-mix nebulizer-burner system is shown in exploded view in Figure 48. The sample solution is drawn into the nebulizer by the reduced pressure created at the end of the capillary by the oxidant or carrier gas flow. The resulting mist flows, with the carrier gas, into the spray chamber where the droplets are homogenized. Big droplets (diameter >10 μm) will fall onto the sides of the chamber and flow to waste. When a laminar flow burner is used, the flame is very stable and the sample mist only has a small affect on the characteristics of the flame. Two exceptions of this statement are when organic solvents are employed, or when the volume ratio of the sample solution and combustion gases exceeds a critical ratio of about 1:5000. The organic

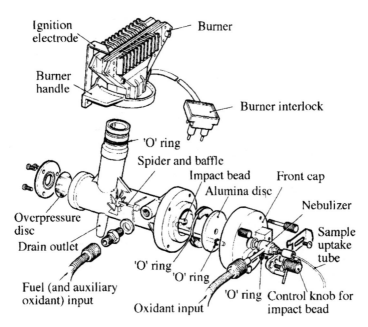

Figure 48 *Premix burner*
(Unicam Analytical Systems Ltd.)

solvents alter the carbon-oxygen ratio in the flame. Despite the complexity of the pre-mix system, its response time is fast. It only takes about one second from the introduction of the sample solution until the signal reaches the readout device. A steady reading is obtained in 7 to 10 seconds, when dynamic equilibrium in the flame is reached.

The indirect injection burner is safer than the pre-mix burner, because the fuel gas and oxidant are mixed at the orifice of the burner and no danger from explosion hence exists. There is some turbulence in the flame, because the gases are mixed and burned in the same region. In addition, the flame is loud. The direct injection burner is especially used for determinations in organic solvents. A quartz or ceramic absorption tube is frequently used in connection with the direct injection burner (Figure 49).

3.2.2.1 Pneumatic Nebulizer. A pneumatic nebulizer unit is shown in Figure 50. These systems are the most widely used nebulizers to produce mists of the sample solution in AAS. The efficiency and the droplet size distribution mostly depend on the diameter and the relative positions of the end of the capillary and the nose-piece. The effect of the sample uptake rate on the absorbance is shown in Figure 51. A maximum

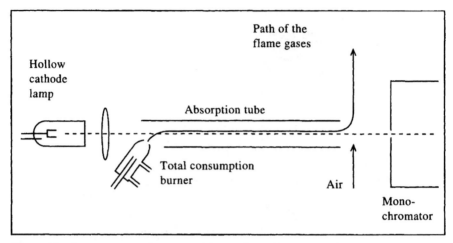

Figure 49 *Direct injection burner together with an absorption tube*

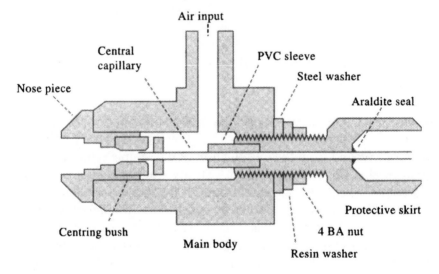

Figure 50 *Pneumatic nebulizer*
(Unicam Analytical Systems Ltd.)

is obtained between flow rates of 3 and 6 ml min^{-1} when the efficiency is about 10%.

The diameter of the droplets produced by a pneumatic nebulizer varies from <5 μm to about 25 μm. The spray chamber allows droplets to reach the burner which can be vaporized and atomized in the flame. If the spray chamber prevents small droplets (diameter of about 10 μm or less) from entering the flame, sensitivity will be decreased. On the

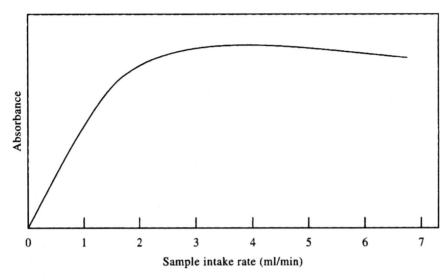

Figure 51 *Effect of the uptake rate on the measured absorption signal*

other hand, if large droplets (>10 μm) reach the flame, the flame noise will increase and the temperature will decrease. From the total mass of the sample nebulized, the maximum useful amount of droplets is about 10%, which gives the limit for the maximum attainable efficiency of the nebulizer. However, the nebulization efficiency can be improved in various ways by altering the droplet size distribution. A bead or bar placed close to the orifice of the nebulizer, or a counter flow nebulizer can be used for this purpose (Figure 52).

The droplets are hitting the impact device with nearly sonic speed resulting in fragmentation which can increase the mass of material vaporized by 50 to 100%. The mechanical device must be chemically inert (stainless steel coated with inert plastic). In the counter flow

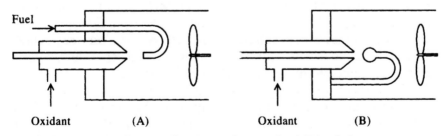

Figure 52 *Principle of counter flow (A) and impact bead (B) nebulizers*

nebulizer, the oxidant-sample aerosol and the fuel gas nozzles are placed in opposition within the spray chamber.

3.2.2.2   Spray Chambers.   The selection of sample mist of the desired droplet size and the complete mixing of the sample mist with the oxidant and fuel gases before entering the burner occurs in the spray chamber. The fuel gas enters the spray chamber tangentially. Auxiliary air or dinitrogen oxide is also introduced into the spray chamber. Auxiliary oxidant is often required to support the flame.

Some vaporization may occur in the spray chamber, and the largest droplets will condense onto the walls of the chamber and go to waste. The waste tube leads to a liquid trap which prevents the gases from escaping and ensures a small steady excess of pressure in the spray chamber. Because a relatively large volume of flammable gas is in the chamber, it is a potential source of danger. Modern AA instruments, however, are equipped with gas control systems which give protection from flashback of the flame. Spoilers are often employed inside the spray chamber to improve the change between the sample mist and the tube walls.

3.2.2.3   Ultrasonic Nebulizer.   The ultrasonic devices are more efficient nebulization systems than the pneumatic nebulizers, but they possess a lower nebulization rate. By the use of these systems, it is possible to convert about 40% of small sample volumes to a mist of useful droplet size, but sensitivity is generally not increased because of the slowness of the nebulization.

Ultrasonic nebulizers may divide into the liquid coupled and vertical crystal systems. Ultrasound is produced by the piezo-electric effect on certain crystals. The crystal vibrations are then transmitted to the solution to be nebulized.

The principle of the liquid coupled ultrasonic nebulizer is shown in Figure 53. The sample solution is held in contact with the coupling liquid (usually water) by a diaphragm. The ultrasonic vibrations are focused by a concave crystal through the coupling medium to the sample. The sample mist formed is swept away to the burner by a carrier gas (air). Frequencies of 70 and 115 kHz have been used, and the average droplet size produced is about 17 to 20 μm diameter range. Only small samples can be handled with these systems, but they cannot be fed in continuously as with the pneumatic nebulizers. This problem may be overcome by the vertical nebulization devices.

In the vertical ultrasonic nebulizer the sample solution is injected continuously at a constant rate directly onto the vertical face of the vibrating quartz crystal (Figure 54). The nebulization efficiency is about 50 and 25% at sample flow rates of 0.1 and 3 ml min^{-1}, respectively.

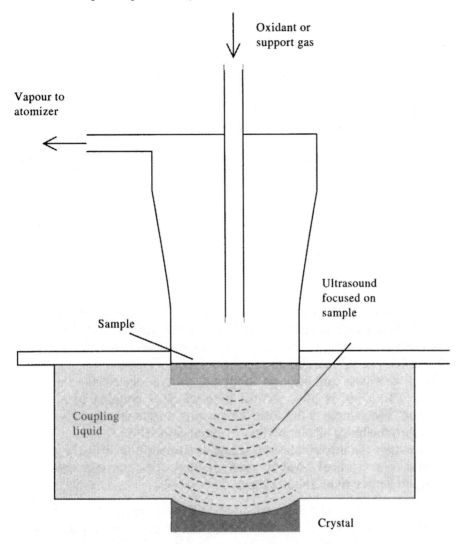

Figure 53 *Principle of the liquid coupled ultrasonic nebulizer*

The stability and reproducibility of the absorbance readings obtained with pneumatic nebulizers are better than with ultrasonic nebulizers.

3.2.2.4 Direct Introduction of Solid Samples. Powdered solids can be introduced directly into the flame with graphite capsules and vibration tubes, and as suspensions.

A solid sample of about 50 mg is placed in a porous graphite capsule. The ends of the capsule are closed with graphite powder and the capsule is placed horizontally in the flame. Atomic vapour will then diffuse

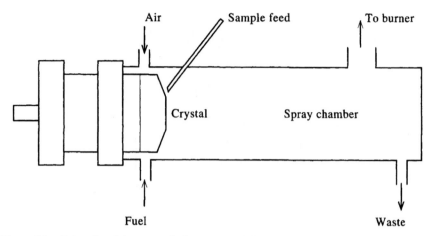

Figure 54   *Principle of the vertical ultrasonic nebulizer*

through the graphite tube and its absorption will be measured above the capsule. Small solid particles cannot diffuse through the graphite and in this method no background scatter will exist. Detection limits between 10 and 100 mg kg^{-1} can be obtained for many elements.

Solid powders can also be analysed with a pneumatic sampling device. This consists of a vertical tube which is coupled to a 50 Hz vibrator. The sample is fed into the air inlet nozzle at a constant rate. The reproducibility of the method is about 8 to 10%.

Nowadays the introduction of solids as suspensions of finely ground powder is a relatively widely used method. The particle size must generally be less than about 12 µm.

3.2.2.5   Burners.   The main requirements for commercial burners are the maintenance of a large safety factor and ability to pass solutions of highly dissolved solids without blockage. Usually acidic solutions are prepared for the AA analysis, which means that burners are made from stainless steel or titanium in order to avoid corrosion. Also the inner surfaces of the burner shell are coated with an inert plastic material, like poly(phenyl sulfide).

The construction of the burner depends on the oxidant/fuel gas mixture. The flame propagation velocity should be smaller than the gas flow velocity through the burner slot. In the opposite case, the flame may flash back down the burner stem and into the spray chamber with possible disastrous results. The length of the slot in air-acetylene, air-propane, and multislot burners is 10 cm, and that of the dinitrogen oxide-acetylene burner is 5 cm.

The multislot or Boling burner gives a wide atom vapour which completely surrounds the radiation signal from the radiation source. Thus, the outer mantel of the flame protects the inner part of the flame against atmospheric air. The multislot burners are employed with the Delves-cup and boat methods (so called semiflame methods).

In separated burners, the flame is separated on both sides from atmospheric air with an inert gas flow (nitrogen or argon). These burners are employed for determination of refractory oxide elements and when atmospheric air causes interference to the measurement.

3.2.3 Atomization Process in the Flame

In the atomic absorption techniques the vapour of free, unionized atoms of the analyte element is led to the optical path of the light beam originating from the radiation source. The success of the analysis depends primarily on the production of the uncombined and unionized atoms in the ground state.

As the sample solution, in which the analyte element is in the form of a dissolved salt (for example, $KCl(s) \rightarrow K^+ (aq) + Cl^- (aq)$), enters the flame at a temperature of 2000 to 3000 K, the atomization process is considered to occur as follows: (i) First the solvent rapidly evaporates and a solid aerosol will be formed ($K^+(aq) + Cl^-(aq) \rightarrow KCl(s)$); (ii) Then the solid particles melt and vaporize ($KCl(s) \rightarrow KCl(g)$). The vaporization is fast, provided the melting and boiling points of the analyte compound are well below the temperature of the flame; (iii) The vapour consists of separate molecules or molecule aggregates which tend to decompose into individual atoms ($KCl(g) \rightarrow K(g) + Cl(g)$); (iv) Individual atoms may absorb energy by collision and become excited or ionized ($K(g) \xrightarrow{collision} K^*(g)$ or $K(g) \xrightarrow{collision} K^+(g) + e^-$); (v) Free atoms in the ground state can absorb radiation energy ($K(g) + h\nu \rightarrow K^*(g)$). To obtain the best available sensitivity, the dissociation of the sample to neutral atoms should be complete and no excitation or ionization by collision should happen. The excitation due to the high temperature of the flame is negligible, as shown before.

The distance which the sample travels through the flame before atomization occurs is very short. For example, in the case of sodium chloride this distance is less than 1 cm in the air-acetylene flame (Figure 55). In the higher region of the flame the sodium atom concentration will slowly decrease due to cooling of the flame. The composition of the flame has little affect on the atomization of sodium, while refractory oxide elements behave differently in rich and lean flames (Figure 55). In

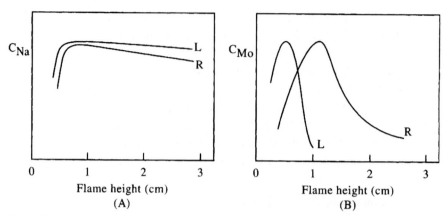

Figure 55   *(A) Na atom concentration versus air-acetylene flame height; (B) Mo atom concentration versus air-acetylene flame height (R = rich and L = lean)*

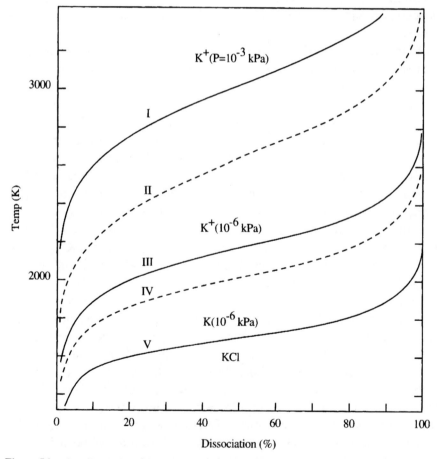

Figure 56   *Atomization and ionization curves for potassium*

lean flames the higher oxygen concentration will favour the formation of thermally stable molybdenum oxide.

Dissociation and subsequent ionization of a diatomic compound are by nature similar processes. It is possible to calculate by means of the Saha equation (Equation 62) degrees of ionization of atoms into ions and electrons for different elements as a function of temperature and partial pressures. Figure 56 shows atomization and ionization curves for potassium as a function of temperature at different partial pressures.

The partial pressure of potassium atoms under conventional AA measurement conditions (the sample uptake rate 3 ml min^{-1}, and the acetylene and air flow rates of 1.5 and 6 l min^{-1}, respectively) is about 10^{-4} kPa when the potassium concentration in the sample solution is 100 mg l^{-1}. The corresponding pressure for a solution containing 1 mg l^{-1} of potassium is 10^{-6} kPa.

The Saha equation is:

$$\log K = -U/4.573T + 2.5 \log T - 6.49 + \log(g_{K^+} \times g_e)/g_K \qquad (62)$$

where K is the equilibrium constant for the ionization process of potassium atoms $(K(g) \to K^+(g) + e^-)$, g_i is the statistical weight of the species i, U is the ionization energy (cal mol^{-1}), and T is the absolute temperature (K). For potassium the equation reduces to:

$$\log K = -21873/T + 2.5 \log T - 6.49 \qquad (63)$$

This equation can be used to calculate the equilibrium constants at different temperatures between 1800 and 3000 K. The degrees of ionization at various partial pressures can then be calculated using the equation:

$$K = (p_{K^+} \times p_e)/p_K = (Px^2)/P(1-x) = Px^2/(1-x) \qquad (64)$$

where p_{K^+}, p_e, and p_K are the partial pressures of the species shown as suffixes, P is the total pressure of potassium atoms and ions, and x is the degree of ionization $(0 \le x \le 1)$.

Figure 56 shows the calculated degrees of ionization of potassium atoms as a function of temperature for solutions in which the potassium concentration is 100 mg l^{-1} (curve II) and 1 mg l^{-1} (curve III).

Degrees of dissociation of potassium chloride can be derived in an analogous way to the ionization of potassium atoms (Equation 56). Curves IV (100 mg l^{-1}) and V (1 mg l^{-1}) represent dissociation of KCl.

Relative amounts of KCI, K, and K^+ species for various pressures and concentrations at a given temperature are obtainable from the curves. This presentation is very ideal and gives only an approximation to the real situation.

3.2.4 Interferences in Flame Atomization

Atomic spectrometric methods are relative methods. The signals produced by the sample solutions are compared to the signals caused by the reference solutions. If the samples and references are behaving differently during the measurements, interferences will be seen. Interferences in the flame AA techniques may be divided into chemical, ionization, physical, and background absorption interferences.

3.2.4.1 Chemical Interferences. Chemical effects originate either in the flame or in the sample solution. The interference mechanisms may be divided into the two groups: (i) The atomization of the analyte element is not completed either in the solid phase or in the liquid (condensed) phase; (ii) The vaporized atoms react with other atoms or radicals present in the gas phase.

Chemical interference will occur when the analyte element forms with another element, radical, or compound a new compound in the condensed phase and this new compound possesses different thermochemical characteristics. The interference becomes greater with increasing difference in the dissociation temperatures of the original and new compounds. Dissociation is dependent on the flame temperature, the ratio of oxidant/fuel gas, the concentration of the analyte element, the efficiency of the nebulizer, and the measurement height in the flame.

Stable compound formation will always cause a depressive effect. Typical examples are the lowering of alkaline earth metal absorbances in the presence of phosphate, aluminate, silicate and some other oxo anions, the low sensitivity of metals which form thermally stable oxides (refractory oxide elements), and the depression of the calcium signal in the presence of proteins. In addition, some refractory oxide elements may also form stable carbides, especially in rich hydrocarbon flames.

Among the most common chemical interferences in flame spectrometric methods, is the signal depression of alkaline earth metals in the presence of phosphate. This phenomenon occurs both in absorption and emission measurements. The interference effect is due to the formation of solid alkaline earth metal pyrophosphates which are difficult to vaporize. If calcium and phosphate solutions are simultaneously aspirated into the flame with two different nebulizers, signal depression is not observed. This is an indication that calcium phosphate does not form in the flame, but at an earlier stage of the process. The effect of phosphate can be removed by adding large amounts of lanthanum into the solutions to be analysed. The determination can be carried out without interference, when the lanthanum concentration is about five times larger than that of the phosphate. Interference caused by phosphate, sulphate, aluminate, and silicate anions in the determination of alkaline earth metals, has also been reduced by addition of strontium, neodymium, samarium, and yttrium salts. Organic chelating agents which form stable complexes with the analyte, can also be used for counteracting chemical interferences of various anions. For example, EDTA or glycerol may be used to eliminate interference caused by phosphate and sulphate in the determination of aluminium.

From the interferences caused by cations, the interferences between aluminium and alkaline earths are common. For instance, the determination of magnesium in samples containing large amounts of aluminium (for example, silicate minerals) is difficult. The interference is caused by the formation of a thermally stable mixed metal oxide, $MgAl_2O_4$ (spinell). This case is also dependent on the acid used and its concentration (Figure 57). The different influences of HCl in

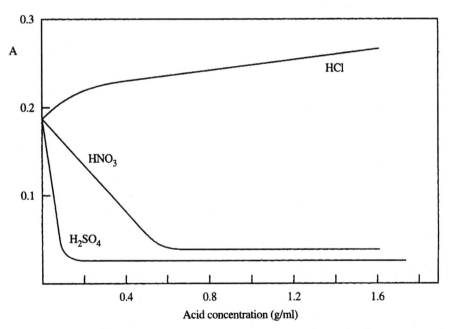

Figure 57 *The effect of aluminium (2 g l^{-1}) and some acids on the determination of magnesium (5 mg l^{-1})*
(Adapted from: W. T. Elwell and J. A. E. Gidley, Atomic Spectrometry, 1966)

comparison with HNO_3 and H_2SO_4 have been attributed to the different thermal behaviour of the corresponding metal salts. Metal nitrates and sulphates are decomposed directly to the corresponding oxides, whereas metal chlorides decompose first to free atoms which then can be oxidized. The influence of aluminium on the determination of calcium and magnesium can be avoided with addition of strontium to the sample and standard solutions.

Various methods of counteracting interference by stable compound formation in the condensed phase are available: (i) Removal of the interfering anion or anions by separation techniques (*i.e.* ion exchange, liquid-liquid extraction); (ii) Additions of an excess of the interfering anion to both sample and standard solutions; (iii) Use of hotter flames; (iv) Addition of the salt of a metal which will form a stable compound with the interfering anion.

The reagents used for counteracting chemical interferences may be divided into 'releasing' and 'shielding' agents. A releasing agent, forms a thermally less stable compound with the analyte than the interfering species. Shielding agents form metal complexes either with the analyte or interferent and by preventing the reaction between the analyte and interferent allow them to vaporize separately. Table 8 lists chemical interferences for some elements with the proposed releasing or shielding agents.

Interferences in the condensed phase may also be minimized by reducing particle size or delay time of the sample in the flame. The particle size must be as small as possible so that the evaporation would be fast. Drop size may be controlled by the sample intake rate, use of the impact beads, or counter flow nebulizers, and organic solvents.

Table 8 *Chemical interferences in the determination of some elements, and reagents for eliminating the interferences (R = releasing, S = Shielding)* (Source: T. C. Rains, 'Chemical Interferences in Condensed Phase, in Flame Emission and Atomic Absorption Spectrometry, Vol. 1', eds. J.A. Dean and T. C. Rains, Marcel Dekker, New York, 1969)

Analyte	Interfering element/ion	Reagent/type
Mg, Ca	Al, Si, PO_4^{3-}, SO_4^{2-}	La/R
Na, K, Mg	Al, Fe	Ba/R
Sr	Al, PO_4^{3-}, SO_4^{2-}	Nd, Sm, Y, Pr/R
Mg, Ca, Sr, Ba	Al, Fe, Th, lanthanides, Si, B, Cr, Ti, PO_4^{3-}, SO_4^{2-}	glycerol, $HClO_4$/S
Na, Cr	Al	NH_4Cl/S
Mo	Sr, Ca, Ba, PO_4^{3-}, SO_4^{2-}	NH_4Cl/S
Ca	PO_4^{3-}	mannitol, ethylene glycol/S
Mg, Ca	Se, Te, Al, Si, NO_3^-, PO_4^{3-}, SO_4^{2-}	EDTA/S

Interferences in the solid phase will increase with decreasing temperature of the flame. For example, in the air-propane and air-hydrogen flames the thermal dissociation is incomplete. The analyte may react in the flame with oxygen, hydroxyl radicals, or hydrogen to form new compounds, of which oxides are the most important. The oxide formation can be minimized in reducing conditions by increasing the carrier flow rate with respect to the oxidant flow rate.

The observation point in the flame may also affect the absorption signal. If the analyte forms a thermally stable oxide, the degree of the interference may vary in different positions of the flame.

The sensitivity is directly proportional to the dissociation constant of the analyte compound. The composition of the flame must be constant during the measurements in order to obtain good precision and accuracy.

3.2.4.2 Ionization Interferences. The analyte is partly ionized in hot flames leading to decreased absorption signals. This type of interference is important for elements with low ionization potentials (alkali metals and some alkaline earth metals). For example, a significant amount of potassium atoms are ionized at 2000 K (Figure 56).

Ionization effects can be removed or reduced by adding a large excess of an element which ionizes more easily than the analyte. The added element acts as an ionization buffer by reducing the ionization of the analyte. Ionization of potassium can be removed by lithium, although the higher alkali metals (with lower ionization potentials) are even more efficient. The effect of potassium on the determination of calcium, strontium, and barium is shown in Figure 58.

3.2.4.3 Physical Interferences. The physical interferences originate in changing physical characteristics of the solutions to be measured (viscosity, surface tension, vapour pressure, temperature). Various solution properties have an effect on the sample intake, nebulization, transport of the sample mist, evaporation of the solvent, vaporization of the analyte, and scatter of the light.

In practice, very dilute solutions are used in AAS, so that the physical properties of the sample solutions are close to those of the pure solvent (water). Physical properties will be changed if the sample solution contains large amounts of acids, salts, or organic compounds. Concentrated solutions are viscous which may cause problems with the sample intake and nebulization. The absorption signal decreases with increasing viscosity. Viscosity interferences may be avoided by using the same solvent for samples and standards and by matching the acid and salt contents in each solution to be measured. In practice, the total

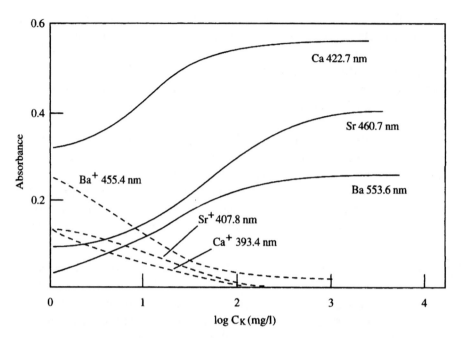

Figure 58   *The effect of potassium on the determination of calcium (5 mg l⁻¹), strontium (5.5 mg l⁻¹), and barium (30 mg l⁻¹) using dinitrogen oxide-acetylene flame* (Adapted from: M. D. Amos and J. B. Willis, Spectrochim. Acta, 1966, **22**, 1325)

salt content should be less than 0.5%. For very difficult samples, it is advantageous to use sample-like standards and a method of standard additions.

Nebulization is clearly dependent on the surface tension. The surface tension of the organic solvents is smaller than that of water and, hence, in organic solutions the droplets of the sample mist are smaller than in aqueous solutions. The vaporization will be improved if the vapour pressure of the organic solvent is greater than that of water. This is why the sensitivity in organic solutions is better than in aqueous solutions.

3.2.4.4   Spectral Interferences.   Relatively few examples of actual spectral interferences have been reported in atomic absorption or atomic fluorescence spectrometry. This means that the possibility that a resonance line emitted from a line-like radiation source may overlap with an absorption line of another element present in the atomizer is very small.

In AAS spectral interferences may be classified into the four groups: (i) More than one absorption line in the spectral bandpass; (ii)

Non-absorbed line emitted by excitation source; (iii) Spectral overlap in atom source; (iv) Continuum or broad band absorption and scatter.

First case has a calibration graph with two linear segments, and then the graph becomes asymptotic to the transmission value of unabsorbed light.

Second case would normally be referred to as a band-pass effect. However the resonance and offending non-absorbed lines which are sufficiently close to pass through the exit slit of the atomic absorption monochromator would be well separated by a higher resolution monochromator. The unabsorbable radiation passing through the exit slit gives rise to the more conventional type of calibration curvature defined by the stray light equation.

In the third case the interfering line is sufficiently close to the wavelength of the absorption line of the analyte to absorb some of the radiation of the emission line, leading to an erroneously high result. The spectral overlap is only overcome by removing either the interfering element or the element to be determined from the sample or, if possible, by using another absorption line for the analysis.

A number of cases of spectral overlap have been reported, and these are summarized in Table 9. It can be noted that atomic absorption and emission profiles are not always as narrow as is usually expected, and the spectral interference may be experienced at line separations even greater than 0.04 nm.

The effects of the fourth case will be seen if the absorption band of a molecule existing in the flame is located at the same wavelength as the resonance line of the analyte. Molecular absorption spectra obtained from several alkali and alkaline earth halides are shown in Figure 59. The effects of continuum band absorption and scatter are discussed in detail in the sections on Background Correction (Section 3.4).

3.3 ELECTROTHERMAL ATOMIZATION

Electrothermal atomization can be performed by graphite furnace, open filament, or vertical crucible furnace devices. Graphite furnace atomizers are the most widely used of these techniques.

3.3.1 Graphite Furnace Atomizers

The use of a graphite furnace for generating free atoms in atomic absorption spectrometry was first introduced by L'vov in 1959. In the original L'vov furnace the sample to be analysed was placed on a sample electrode, which was then introduced vertically into the

Table 9   *Spectral overlaps in AAS*
             (Source: J. D. Norris and T. S. West, *Anal. Chem.*, 1974, **46**, 1423
             and references therein)

Source[a]	Emission wavelength (nm)	Analyte	Absorption line (nm)	Separation (nm)
Al	308.215	V	308.211	0.004
Sb	217.023	Pb	216.999	0.024
Sb	217.919	Cu	217.894	0.025
Sb	213.147*	Ni	231.095	0.052
Sb	323.252	Li	323.261	0.009
As	228.812	Cd	228.802	0.010
Cu	324.754*	Eu	324.753	0.001
Ga	403.298*	Mn	403.307	0.009
Ge	422.657	Ca	422.673	0.016
I	206.163	Bi	206.170	0.007
Fe	271.903*	Pt	271.904	0.001
Fe	279.470	Mn	279.482	0.012
Fe	285.213	Mg	285.213	<0.001
Fe	287.417*	Ga	287.424	0.007
Fe	324.728	Cu	324.754	0.026
Fe	327.445	Cu	327.396	0.049
Fe	338.241	Ag	338.289	0.048
Fe	352.424	Ni	352.454	0.030
Fe	396.114	Al	396.153	0.039
Fe	460.765	Sr	460.733	0.032
Pb	241.173	Co	241.162	0.011
Pb	247.638	Pd	247.643	0.005
Mn	403.307*	Ga	403.298	0.009
Hg	253.652*	Co	253.649	0.003
Hg	285.242	Mg	285.213	0.029
Hg	359.348	Cr	359.349	0.001
Ne	359.352	Cr	359.349	0.003
Si	250.690*	V	250.690	<0.001
Zn	213.856*	Fe	213.859	0.003

[a]Radiation sources were hollow cathode lamps except Ne, I, As, and Hg (EDLs); *Resonance line from the source.

horizontal graphite tube through a hole in the underside of the tube (Figure 60). The graphite cuvette was 5 to 10 cm in length and 3 mm in diameter. By this method it was possible to analyse both liquid and solid samples. In the case of liquid samples, solvent must be evaporated before atomization. In early devices the atomization temperature needed was generated first with an arc between the sample electrode and graphite tube, and then by an electric current flowing through the electrode and sample. To prevent oxidation of graphite, the tube and electrode were held within a water-cooled glass casing which contained an inert atmosphere.

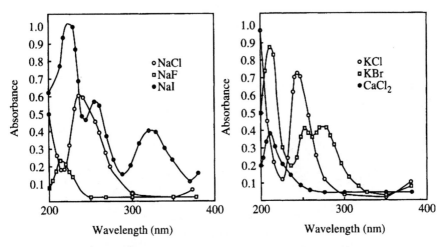

Figure 59 *Molecular spectra of NaCl, NaF, NaI, KCl, KBr, and CaCl₂*
(Adapted from: B. R. Culves and T. Surles, Anal. Chem., 1975, **47**, 920)

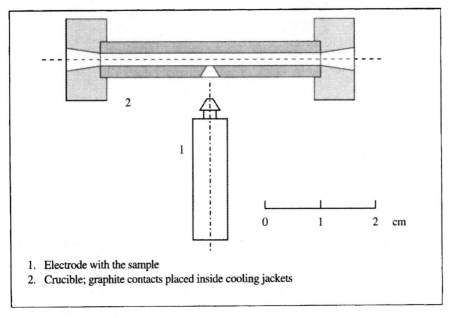

1. Electrode with the sample
2. Crucible; graphite contacts placed inside cooling jackets

Figure 60 *L'vov furnace*

In the Massmann's furnace, published in 1968, there is no sample electrode and the tube dimensions were about 5 cm in length and 6.5 mm in diameter (Figure 61). Liquid samples of 2 to 200 µl volume were introduced through a 2 mm hole placed halfway along the tube. The tube was placed between two graphite end-cones and the whole

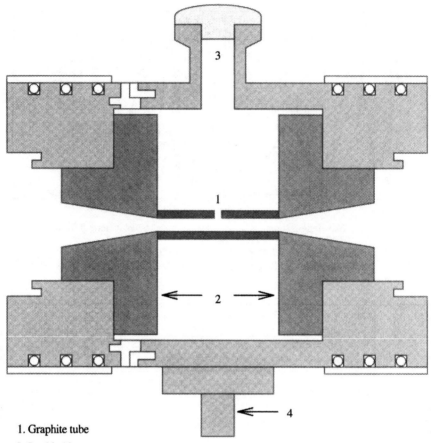

1. Graphite tube
2. Steel holders
3. Sample inlet port
4. Mounting holder

Figure 61   *The Massmann furnace*

device was housed inside a stainless steel water-cooled jacket which contained an inert atmosphere to prevent oxidation of the graphite. Source radiation passes through quartz windows in the jacket. The tube was heated electrically with a low voltage supply connected to the graphite cones. The operating cycle of this model was based upon three principal steps: (i) The drying step to remove the solvent; (ii) The pyrolysis or ashing step to remove as much as possible of the matrix material; (iii) The atomization step to produce free atoms of the element to be determined.

Wooddriffs furnace (introduced also in 1968) was operated at a constant temperature without a temperature programme. Its construction was much larger than that of the Massmann's furnace. Liquid samples are introduced in a nebulized form in a stream of an inert gas. Solid samples were introduced into the atomizer in a small graphite boat.

At the same time as Massmann and Wooddriff other research workers such as West and Williams (1969) and Alder and West (1970) were developing ohmically heated tungsten and graphite filaments and rods for electrothermal atomization. However, over the years these devices have fallen out of favour and at the present time all commercially available modern electrothermal atomizers are different variations of the original Massmann graphite furnace design.

3.3.1.1 Stabilized Temperature Platform Furnace. The conditions in a Massmann furnace are by no means optimum during atomization. The sample is dispensed onto a cold wall of the tube which is then heated rapidly for atomization. After vaporization the sample is in an environment that is not in equilibrium with respect to temperature or time. This can cause chemical interferences. Fewer interferences would occur in a continuously heated, isothermal furnace. It has been shown that about 60% of the atoms formed diffuse to the cooler cuvette ends and condense. In addition, at 2800 K a temperature gradient of 1000 K between the middle and the ends of a graphite cuvette has been measured. The equilibrium with respect to time is even more important than the temperature constancy over the length of the cuvette.

L'vov and his co-workers proposed that the sample should not be introduced onto the tube wall, but onto a 'platform' mounted in the tube (Figure 62). The platform is placed in the tube so that there is virtually no heat transfer through direct contact. If the graphite tube is heated rapidly in the atomization step, the temperature of the platform follows this temperature rise sluggishly. When the tube and the gas phase have reached the atomization temperature and are largely in equilibrium, the platform is heated rapidly by the radiation and the hot gas. This means that the sample is atomized into an environment in which the temperature is not changing (Figure 63), and fewer chemical interferences will be found.

Perkin Elmer introduced, in 1984, a 'Stabilized Temperature Platform Furnace (STPF)' (Figure 64). There are two inert gas flows (internal and external). The heating rate of the furnace is high (about 2000 K s^{-1}), which allows the use of lower atomization temperatures. After the thermal pretreatment a small temperature step (<1600 K) is used in order to prevent the atomization of the sample before the equilibrium has been reached.

In the atomization step the internal gas flow is stopped. The closed structure of the furnace ensures the stable gas phase inside the tube,

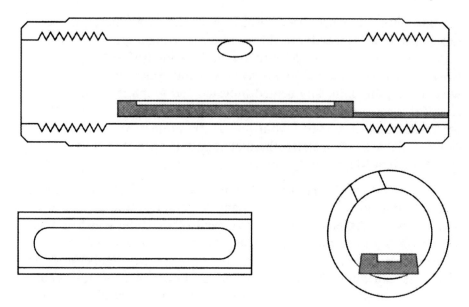

Figure 62   *Graphite tube with L'vov platform*
(Perkin Elmer Corp.)

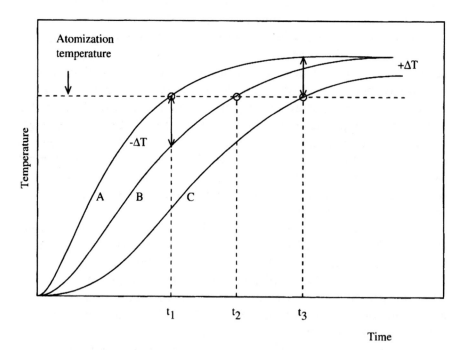

Figure 63   *Temperatures of graphite tube wall (A), gas (B), and L'vov platform versus time*

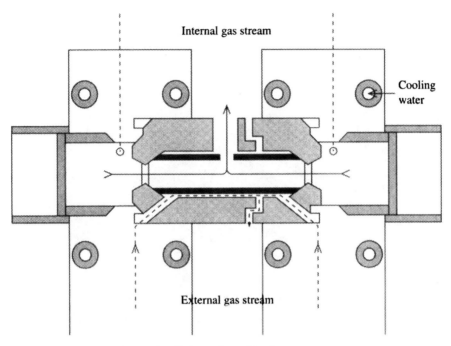

Figure 64 *Modern longitudinally heated graphite furnace*
(HGA-500, Perkin Elmer Corp.)

and prevents the cooling and removal of the atom vapour during the atomization.

The latest graphite furnace design is the transverse heated graphite furnace with an integrated L'vov platform (Figure 65). The tube and platform are machined from one piece of graphite. This new design provides a uniform temperature profile over the entire tube length, which means that the tube ends reach the same temperature as the tube centre (Figure 66). The formation of free atoms is optimum, and recombination of atoms to molecules, the loss of atoms, and condensation on cooler tube ends are effectively avoided.

Memory effects caused by the condensation of the matrix and analyte on cooler graphite parts with a conventional, longitudinally heated graphite tube are minimized in the transverse heated graphite tube (Figure 67).

The following benefits can be obtained by the transverse heated graphite tube and rapid heating: (i) STPF conditions for interference-free analysis for all elements including the refractory elements like V, Ti, and Mo; (ii) Time saving and simplified operating since no tube change is required for the determination of refractory elements; (iii) Automatic multi-element analysis of significantly different elements is

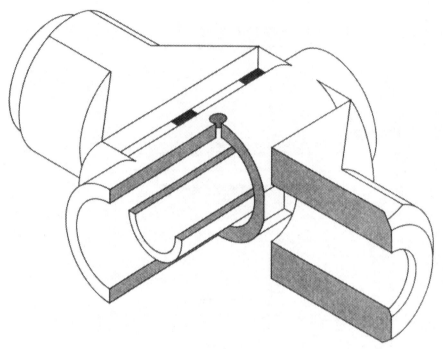

Figure 65   *Cross-section of the transverse heated graphite tube with integrated L'vov platform*
(4100 ZL graphite tube, Perkin Elmer Corp.)

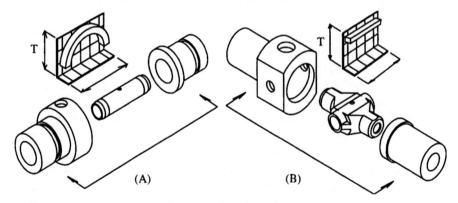

Figure 66   *Comparison of principles of the longitudinally heated graphite tube (A) and the transverse heated graphite tube (B)*
(Perkin Elmer Corp.)

possible; (iv) Improved day-to-day precision by using a fixed platform; (v) Longer graphite tube lifetimes and shorter analysis programmes since the rapid heating of the low mass platform leads to maximum atomization efficiency and to much lower atomization temperatures.

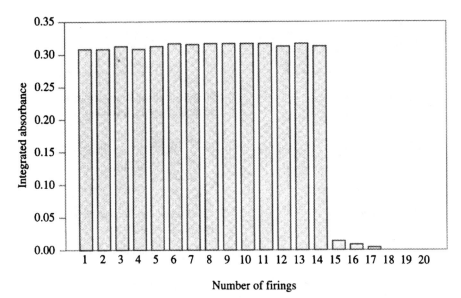

Figure 67 *Examples for reduced memory effects by the transverse heated graphite tube in the atomization of Mo (0.4 ng) and 0.2% HNO$_3$ (15–17)* (Perkin Elmer Corp.)

3.3.1.2 Probe Atomization. Probe atomization is an alternative approach to achieve high vapour temperatures in the graphite furnace and isothermal atomization of the sample in order to minimize chemical interference effects in the determination of volatile elements. This technique was first suggested by L'vov in 1978. This modification overcomes some limitations of the platform technique, because in this procedure the heating of the furnace and the volatilization of the sample are separated.

In 1988 Philips Analytical introduced the PU9385X Graphite Autoprobe for probe atomization in GF-AAS. The Autoprobe consists of a mounting holder, a motor, and the graphite probe. Six steps are required for graphite probe atomization as shown in Figure 68: (i) First the sample droplet is injected onto the probe. The probe-head is positioned inside the graphite cuvette beneath the injection hole so that sample deposition can be performed using a conventional furnace autosampler. (ii) In the next step the sample is dried and pyrolysed (ashed) inside the cuvette. The drying and ashing temperatures required are generally higher than in conventional graphite furnace techniques. For example, a drying temperature of 520 to 570 K is needed to dry a 10 µl aqueous sample in 10 to 20 seconds. Pyrolysing or ashing temperatures should always be obtained by performing an ashing plot on the sample matrix studied. (iii) The spectrometer then zeros automatically.

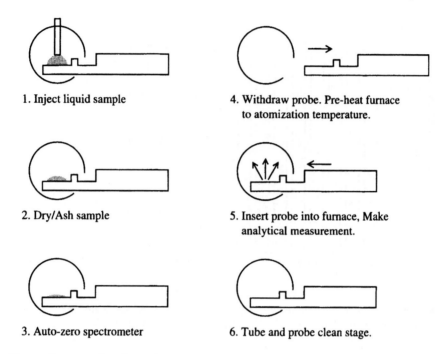

1. Inject liquid sample

2. Dry/Ash sample

3. Auto-zero spectrometer

4. Withdraw probe. Pre-heat furnace to atomization temperature.

5. Insert probe into furnace, Make analytical measurement.

6. Tube and probe clean stage.

Figure 68   *Procedure for probe atomization*
(Unicam Analytical Systems Ltd.)

(iv) In the next step the probe is withdrawn from the atomizer, and the tube is heated to the pre-set atomization temperature. (v) When the cuvette temperature has stabilized, the probe is automatically re-introduced into the furnace. The probe head is heated rapidly by the tube wall radiation and the sample is vaporized into the hot furnace gas. The atomization temperatures required are higher than in conventional graphite furnace atomization. (vi) The final step is a cuvette and probe clean stage.

Two principal advantages of the probe atomization are: (i) Improved control of the vapour temperature the atoms experience; (ii) Rapid heating of the atomization surface to the volatilization temperature of the analyte and matrix. A comparison of typical furnace programmes for wall/platform atomization and probe atomization is given in Table 10. The probe is not in contact with the graphite cuvette. It is heated during the drying, pyrolysing, and atomization stages by convection. Therefore, higher furnace temperatures are required to complete these stages in times comparable to those for wall atomization.

In practice, some drawbacks are observed in probe atomization when compared to the idealized situation described above. The probe

Table 10 *A comparison of typical graphite furnace programmes for wall/
platform atomization and probe atomization*
(Source: D. Littlejohn, Laboratory Practice, 1988, **36**, 121)

Stage	Set tube temperature (K)	Time (s)	Ramp (K s^{-1})
(a) Tube wall/platform atomization (10 µl samples):			
Drying	390 (470)	35 (45)*	20
Thermal pretreatment	570–1470	20	200
Atomization	2270–3270	3	>2000
Cleaning	3070–3270	2	>2000
(b) Probe atomization (10 µl samples):			
Drying	670	45	20
Thermal pretreatment	720–1870	20	200
Pre-atomize	2670–3270	2	>2000
Atomization	2670–3270	3	>2000
Cleaning	3070–3270	2	>2000

*Temperature for platform.

head must be very thin to obtain a low thermal mass to minimize the
cooling effect when the probe is introduced into the cuvette during
atomization stage. However, problems due to weak chemical and
mechanical strength of the probe are observed in routine analysis. In
addition, the spreading of sample droplet on the probe head will limit
the maximum sample volumes available for determination. Also the
sensitivity of probe atomization is lower when compared to platform
atomization. This is due to loss of atoms through the slot cut in the side
of the probe cuvette. The drawbacks mentioned caused the probe
system to vanish from the market during the 90's.

3.3.2 Open Filament Atomizers

In rod atomizers the sample is introduced on a carbon or metal
filament, which is heated electrically in order to atomize the sample.
The design of the West carbon filament atomizer is essentially simple,
consisting of a carbon rod (4 cm in length, 1–2 mm in diameter)
supported between two water-cooled stainless steel electrodes, which are
connected to a low voltage supply. The atomizer is inside a Pyrex glass
chamber through which is flowing an argon stream. Silica windows
allow the radiation beam to pass just above the surface of the filament
for atomic absorption measurements. Rod atomizers can also be
used for atomic fluorescence, but then a side arm at 90° to the optical
axis of the spectrometer is provided for the radiation source. Power
requirements for rod atomizers (about 0.5 kW) are lower than for

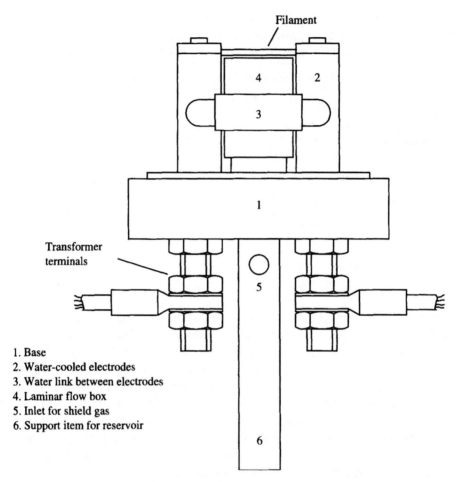

Figure 69   *Rod atomizer or West furnace*
(Adapted from: D. J. Johnson, T. S. West, and R. M. Dagnall, Anal. Chim. Acta, 1973, **67**, 79)

graphite furnaces (several kilowatts). The West atomizer was further simplified by replacing the glass chamber with a vertical argon flow (Figure 69).

The useful sample capacity of open atomizers is about 5 μl. In operation they require a temperature programme (drying, ashing, and atomization steps) like graphite furnace atomizers.

3.3.3   Vertical Crucible Furnace Atomizers

Massmann was the first to publicize a vertically oriented graphite atomizer for atomic absorption and fluorescence measurements

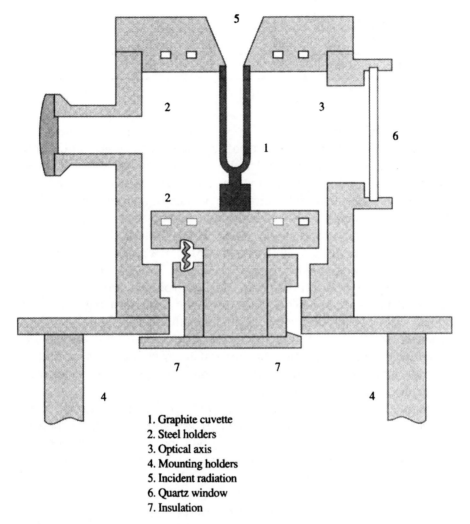

1. Graphite cuvette
2. Steel holders
3. Optical axis
4. Mounting holders
5. Incident radiation
6. Quartz window
7. Insulation

Figure 70 *Vertical crucible furnace*

(Figure 70). This atomizer was designed, especially, for atomic fluorescence spectrometry.

The light beam from the radiation source enters the crucible vertically downwards through the open top and the fluorescence emission is measured horizontally above the sample through a slit in the wall of the atomizer.

3.3.4 Graphite Cuvettes

3.3.4.1 Cuvette Material. An ideal tube material must be chemically inert, have good thermal and electrical conductivity, be obtained in

Table 11 *Typical impurities (µg g⁻¹) in carbon used for graphite tubes*

B	Ca	Cu	Fe	Mg	Si	V	Ash
0.0	0.2	0.1	0.2	0.05	0.2	0.01	2

a state of high purity (Table 11), be machined easily, and have low porosity, a low expansion coefficient, and a high melting point. From the possible materials available, graphite nearly satisfies all these requirements. Tantalum and tungsten have also been considered and used as atomizer materials. Graphite is manufactured from carbon black or petroleum coke by grinding and mixing with binders such as pitches, coal tars, and phenolic resins. Extrusions of the material are made and then heated slowly to about 1600 K in an inert atmosphere. A solid carbon material (amorphous carbon) is formed. It is hard and brittle. Graphite is then made from this amorphous carbon by heating resistively up to about 3300 K. Cuvettes made from normal graphite are named 'electrographite cuvettes'. Graphite sublimes at normal pressures at about 3800 K, which sets the upper temperature limit for its use. Graphite is oxidized at temperatures much lower than this in contact with air, and hence graphite atomizers must always be kept in an inert atmosphere (nitrogen or argon).

Despite the sensitivity of electrographite as the tube material, there are some drawbacks to its use. Normal graphite is naturally porous (porosity about 13%), and the porosity will increase after the graphite has been maintained at high temperatures. Atom vapour tends to diffuse easily through the graphite tube wall, and memory effects will exist. Another disadvantage of normal graphite is that some elements form stable carbides at high temperatures. The melting points of titanium, zirconium, hafnium, niobium, and tantalum carbides are over 3300 K, and those of vanadium, tungsten, molybdenum, and uranium carbides are between 2800 and 3300 K. Calcium, strontium, silicon, aluminium, and boron may also form carbides. These problems may be overcome to a large extent by coating the tubes with pyrolytic graphite or making the tubes entirely of pyrolytic graphite (totally pyrographite cuvettes, TPC).

3.3.4.2 Pyrolytically Coated Graphite Cuvettes. A pyrolytic graphite layer can be formed on an electrographite cuvette substrate by maintaining it in an inert atmosphere containing a small portion of methane or other hydrocarbon at temperatures up to 2800 K. Such conditions lead to the formation of a pyrolytic graphite layer which is

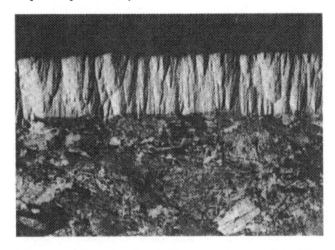

Figure 71 *Photograph of pyrolytic graphite structure on a graphite substrate* (Unicam Analytical Systems Ltd.)

usually 30 to 50 μm in thickness. Figure 71 shows a photomicrograph of pyrolytic graphite structure on an electrographite substrate. The pyrolytic graphite coating is highly ordered, whereas the electrographite has a porous and random structure.

Pyrolytic graphite as an atomizer material has many desirable properties, especially, low gas permeability and good resistance to chemical attack. The reduction in permeability reduces the loss of atomic vapour by diffusion through the cuvette walls and improves residence time, and hence increases the analytical sensitivity. Pyrolytic graphite resists oxidation better than normal graphite by a factor of about ten times. Carbide formation is also reduced in pyrolytic tubes due to the resistance to chemical attack. Hence, carbide forming elements may be determined with improvement in analytical sensitivity and less memory effect. However, in the case of tubes coated with pyrolytic graphite, the coating layer gradually thins as material sublimes until the electrographite substrate is exposed. At this point the analytical sensitivity will decrease similarly to that of an uncoated electrographite tube. Although coated tubes last longer, their performance begins to deteriorate before the tube breaks down (Figure 72).

3.3.4.3 Totally Pyrographite Cuvettes. New types of commercial graphite duvettes which were made entirely of pyrolytic graphite were introduced in 1984 by Philips Analytical. Figure 73 shows a photomicrograph of a cross-section of a totally pyrographite cuvette (TPC). The structure of the cuvette consists of small conical crystallites which radiate from the inner wall surface. These tubes have many superior

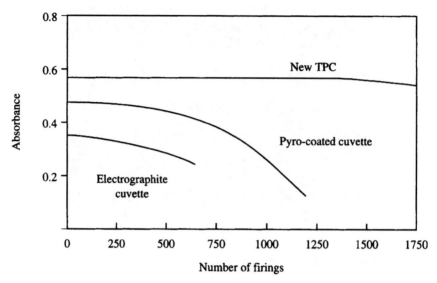

Figure 72 *Comparative performance of TPC cuvettes versus pyro-coated cuvettes and electrographite cuvettes; test element manganese*
(Unicam Analytical Systems Ltd.)

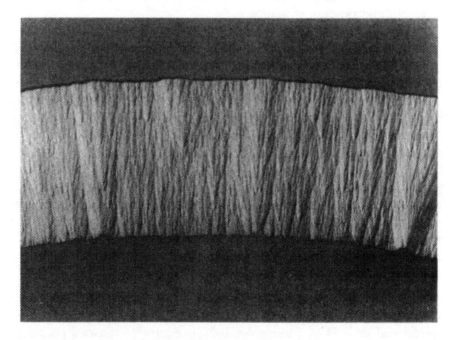

Figure 73 *Photograph of the cross section of a TPC*
(Unicam Analytical Systems Ltd.)

performance characteristics compared to pyrolytically coated tubes. Pyrolytic graphite has a high strength-to-weight ratio (about 5 to 10 times better than electrographite) over the temperature range up to 3300 K. Because of these phenomena, cuvettes can be made with a reduced wall thickness. This gives rise to a lower thermal mass, which in turn means that these cuvettes have increased heating rates (Figure 74).

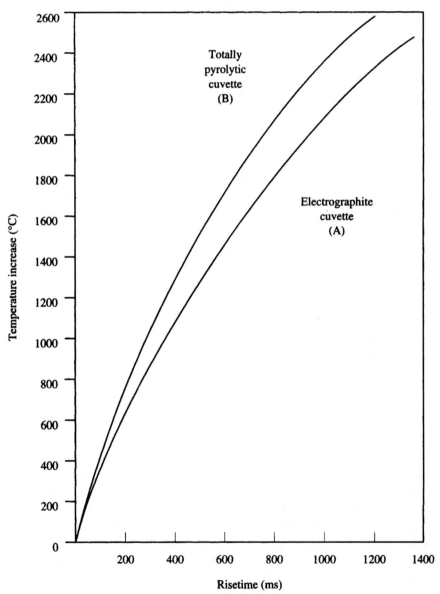

Figure 74 *Temperature risetimes for an electrographite cuvette (A) and TPC (B)* (Unicam Analytical Systems Ltd.)

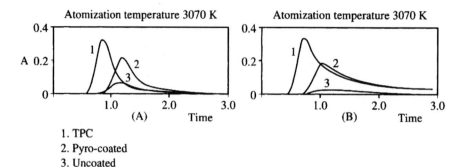

1. TPC
2. Pyro-coated
3. Uncoated

Figure 75 *Peak shapes for platinum (A) and vanadium (B) using TPCs (1), pyrolyti-*
cally coated electrograhite tubes (2), and electrographite tubes (3)
(Unicam Analytical Systems Ltd.)

Pyrolytic graphite is a highly anisotropic material, which means that its physical properties are different in different crystal directions. In a parallel direction to the tube surface the heat conductivity of pyrolytic graphite is about 300 times higher than in the vertical direction to the surface. This means that heating along the tube is rapid, whereas heat conduction outwards is poor.

The lifetime of TPCs is much longer than pyrolytically coated or normal graphite tubes. Figure 75 shows peak shapes for vanadium and platinum using totally pyrographite cuvettes, pyrolytically coated cuvettes, and electrographite cuvettes. The TPC peak is narrower and appears earlier than the peaks of the other cuvette types.

3.3.4.4 Graphite Cuvette Geometries. The first commercial graphite furnace was the Perkin Elmer heated graphite atomizer (HGA 70). The tube dimensions were similar to Massmann's original tube. During the 30 years history of the graphite furnace technique the dimensions of the graphite tubes have been reduced and several new designs have been introduced (Figure 76).

If the tube is large, it will take a long time to attain the highest temperatures and may give rise to greater relative background absorption effects. However, larger sample volumes can be used with large tubes and therefore better analytical precision and relative sensitivity are obtainable. On the other hand, small furnaces tend to have better absorption efficiencies because of their small volumes, as the radiation beam occupies a larger proportion of their measuring volume. Due to this and shorter times to reach the atomization temperature, small furnaces have better absolute sensitivities.

3.3.4.5 Platform Cuvettes. The concept of platform atomization was first described by L'vov in 1978. A platform is a thin rectangular piece

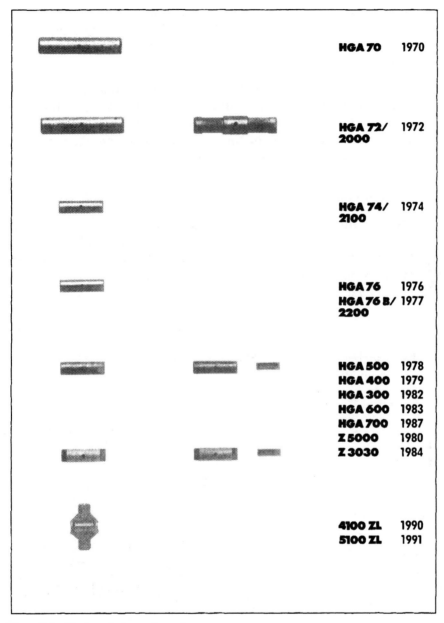

HGA 70   1970

HGA 72/   1972
2000

HGA 74/   1974
2100

HGA 76   1976
HGA 76 B/ 1977
2200

HGA 500   1978
HGA 400   1979
HGA 300   1982
HGA 600   1983
HGA 700   1987
Z 5000   1980
Z 3030   1984

4100 ZL   1990
5100 ZL   1991

Figure 76   *Various designs of graphite cuvettes*
(Perkin Elmer Corp.)

of pyrolytic graphite. The platform is positioned within a conventional graphite cuvette as shown in Figure 62. Cuvettes equipped with integrated platform are also available (Figure 65). Samples are injected on the platform. Contact between the platform and the graphite tube is minimal. Therefore conductive heating from the cuvette wall is negligible, and the platform is heated mainly by the radiation emitted from the inner walls of the graphite tube.

3.3.4.6 Ridged Cuvettes. In ridged graphite cuvettes there are two half rings placed inside the tube on the both side of the injection hole (Figure 77). The rings prevent the sample spreading along the cuvette. These cuvettes are designed for the analysis of samples which are very mobile within the cuvette (especially, organic samples).

3.3.5 Atomization Process

The great sensitivity of electrothermal atomizers is due to their ability to atomize and retain a substantial portion of the analyte in the observation zone for a finite period of time. During the atomization process the rate of formation of the free atoms must be equal or greater than the rate of removal from the optical path:

$$dN(t)/dt = (dN/dt)_{in} - (dN/dt)_{out} \tag{65}$$

At the absorbance maximum the formation and removal rates of the atoms are equal:

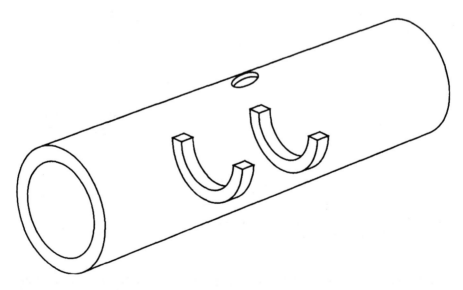

Figure 77 *Ridged graphite cuvette*
(Unicam Analytical Systems Ltd.)

$$(dN/dt)_{in} = (dN/dt)_{out} \qquad (66)$$

It is obvious that the graphite furnace atomizers fulfill this requirement better than the open filament atomizers, which is one of the basic reasons why the former type atomizers are almost exclusively in use nowadays.

The mechanism by which free atoms are produced in the graphite furnace depends on a number of factors, such as the compounds still present in the cuvette at the atomization temperature, the cuvette material, the atmosphere inside the cuvette, the heating rate of the cuvette, and the temperature of operation of the cuvette.

Atoms are removed from the absorption zone by the diffusion process, gas expansion and convection, and recombination. The shape of the absorption signal is determined by the combination of the atom supply and removal functions as shown in Figure 78.

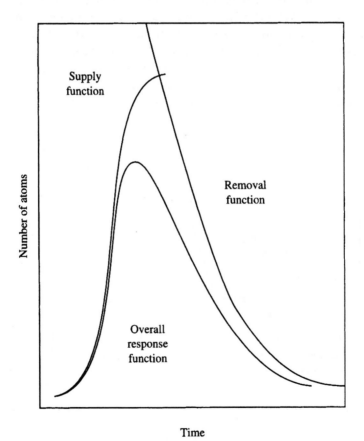

Figure 78 *Response functions and electrothermal atomization*

The process taking place during the thermal pretreatment (ashing) and atomization stages must be known in order to solve the analytical problem with high precision and accuracy. The chemical and physical characteristics of the analyte will determine its behaviour in the furnace. Chemical environment (matrix) of the analyte is also very important. It is possible to predict whether a given element can be determined in a particular matrix and to define the best operating conditions.

In electrothermal atomization the whole sample is atomized at once, whereas in flame atomization the sample is fed continuously into the flame in the form of aerosol. Only about 2 to 3% of the sample reaches the flame atomizer.

The temperature of the graphite furnace can be raised up to about 3300 K. For each analysis a temperature programme must be prepared for the furnace controller. Generally the programme includes the following five basic steps: (i) Drying of the sample; (ii) Thermal decomposition of the matrix (ashing) by raising the temperature up to a level where the analyte remains unatomized; (iii) Producing free atoms of the analyte by rapidly raising the temperature up to the atomization temperature; (iv) Cleaning the furnace by raising the atomizer up to the maximum temperature for a short period of time; (v) Decreasing the temperature down to room temperature using water coolant and inert gas flow (Figure 79). Besides the matrix, the furnace controller

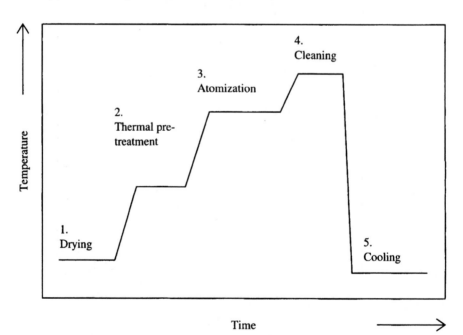

Figure 79   *Temperature programme of the graphite furnace*

programme is dependent on the atomization technique used (tube wall, platform, or probe atomization).

3.3.5.1 Drying. During this step the solvent is evaporated from the sample. Drying must be accomplished in a controlled manner so that the vaporization of the solvent is slow and even. If the solvent is removed by rapid boiling, the sample will froth and splash and some of the sample particles could be carried out of the tube with the gas flow. It must be stressed that the temperature experienced by the sample might be different to that of the programmed furnace temperature. At lower temperatures voltage control is used when the cuvette is electrically heated and a good electrical contact between the cuvette and graphite cones as well as good condition of the grapihite cuvette is essential to obtain a proper temperature. Therefore one has to be aware that the drying of the sample is smooth and even with the used drying temperature. A dentist mirror is a good tool for observing the behavior of sample during the drying phase.

A setting of 380 K is generally high enough for aqueous solutions in tube wall atomization, whereas about 520 to 570 K is needed in probe atomization. The time necessary varies according to the sample size. As an empirical rule, a time in seconds up to 2 times the sample size in µl should be used. Thus, 20 µl of aqueous sample in tube wall atomization needs 30 to 40 s at 380 K to dry. With solvents other than water, the drying temperature and time necessary will obviously be different due to the different boiling points and surface tension.

Organic solvents have low surface tensions and easily wet and soak into the graphite. As they become warm, they tend to spread out over the cuvette. Because of these phenomena, smaller sample volumes and pyrolytically coated cuvettes are generally used with organic solvents.

3.3.5.2 Thermal Pretreatment. In this stage the analyte is separated from the interfering matrix components. Biological samples decompose to carbon and produce lots of soot and smoke. Inorganic compounds distill, sublime, or decompose to mist. If these processes take place at the same time as the atomization of the analyte, the measurement of the absorption signal would be impossible.

When trace metals are to be determined in pure aqueous solutions the thermal pretreatment stage (ashing) has no significance and can be ignored. When trace metals are to be determined in varying kinds of matrix, however, the thermal pretreatment is a very critical stage in the whole furnace programme. The success of the analysis depends on the correct selection of the thermal pretreatment conditions. No general rules can be safely applied since the proper thermal destruction of the sample matrix depends on the matrix itself.

The use of too high a thermal pretreatment temperature or too long a time results in the loss of significant quantities of the analyte before the atomization stage. This is particularly important in the determination of easy volatile elements, such as Hg, As, Se, Cd, Zn, and Pb. The complete removal of the matrix before the atomization stage is possible if the analyte exists in a thermally stable compound.

Thermal pretreatment temperature is very dependent on the matrix and vaporization rate of the analyte. The thermal pretreatment temperature is generally between 470 and 1870 K. For example As, Se and Cd are very easily volatilized, whereas V forms carbide in the graphite furnace and even atomization of this element is very difficult. If the matrix contains several components, two thermal pretreatment steps can be used (it must again be taken into account that the real temperature under voltage control might be different to that of the set temperature).

Three different types of thermal behaviour of the analyte and matrix can be observed: (i) The decomposition of the matrix takes place at lower temperature than the atomization of the analyte; (ii) The decomposition of the matrix takes place at higher temperature than the atomization of the analyte; (iii) The decomposition of the matrix and the atomization of the analyte take place at the same time.

The first case is most common, and if the matrix can be removed completely before the atomization stage, the analysis is free of interferences. In the second case, the determination of easy volatile elements is possible if the atomization of the analyte is complete before the decomposition of the matrix. However, the clean stage must be used in the furnace programme. The third case is typical of easily volatile elements, which can be vaporized partly or completely with the matrix. This problem can be avoided by separating the analyte and interfering matrix from each other before the AAS measurement (liquid-liquid extraction, precipitation, ion exchange, distillation) or using matrix modification. By the use of matrix modifiers the chemical environment of the analyte in the sample is altered. With suitable reagent addition, the evaporation of the analyte will increase or decrease with respect to the matrix.

3.3.5.3 Atomization. Atomization in the graphite furnace can emanate from either molecules or atoms depending on the nature of the sample and behaviour of the analyte. If atomization emanates from molecules, it can be a thermal decomposition or dissociation of a compound, or the reduction of a metal oxide on the hot graphite surface. The difference between these two mechanisms is the active participation of the cuvette material (carbon) in the dissociation of the sample molecules. If atomization emanates from the metal, it can be classified either as desorption or volatilization.

The space required for the analyte is less than 10 mm^2. This area is significantly less than the surface area of the graphite tube. If the analyte element is distributed evenly over the area wetted by the sample solution, then monoatomic or monomolecular layers should exist. Although small heaps of atoms or molecules (crystals) are formed, there is always good contact of the analyte with the tube surface. However, this may not be true when solid samples or highly concentrated salt solutions are introduced into the cuvette.

If atoms are isolated from each other (a monoatomic layer) after drying the sample, atomization will be a desorption from the graphite surface and will be determined by the adsorption isotherm. If the atoms form small heaps or crystals, the atomization will be a pure volatilization. The formation of small atom heaps can be assumed as probable, and thus the volatilization mechanism is more likely.

The atomization of the analyte can occur within 0.1 s, when the saturated vapour pressure is about 10 to 15 Pa. Atomization can begin at considerably lower vapour pressures (temperatures), but then the atomization time is longer. Hence, the time taken to reach the atomization temperature is also very important. With a slow heating rate the atomization takes place slowly, and a considerable portion of the sample is volatilized before the actual atomization temperature is reached. Volatilization will take place in 0.1 s only with a very fast heating rate to the atomization temperature. This is very important since the maximum density of the atom cloud can be achieved only when the atomization time is shorter than the residence time of the atoms in the graphite tube. However, the atomization time in the graphite furnace is still about a thousand times slower than that in the flame. This is also one reason for the relative freedom from chemical interferences in the graphite tube since refractory substances have more time to decompose.

Double curves (thermal pretreatment/atomization curves) are used to determine the limits for both thermal pretreatment and atomization temperatures for the elements and matrices involved. These curves can also be applied to drawing conclusions about the atomization mechanism. In the first curve the absorbance signal (preferably integrated absorbance, i.e. peak area) at the optimum atomization temperature is plotted *versus* the pretreatment temperature as the variable. In the second curve the absorbance at the optimum pretreatment

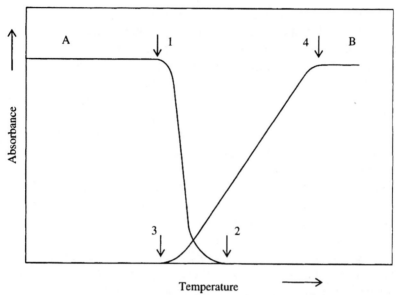

Figure 80   *Thermal pretreatment/atomization curve. A: The absorbance measured at the optimum atomization temperature; B: The absorbance plotted versus the atomization temperature; 1: The maximum thermal pretreatment temperature; 2: The lowest temperature at which the analyte is quantitatively volatilized; 3: The appearance temperature of the analyte; 4: The optimum atomization temperature*

temperature is plotted *versus* the atomization temperature as the variable (Figure 80). The pretreatment curve (ash curve) shows the temperature to which the sample can be heated without loss of the analyte. From this curve it is also possible to derive the lowest temperature at which the element is quantitatively volatilized. From the atomization curve can be derived the temperature at which the atomization is first evident, the appearance temperature, and the optimum atomization temperature at which the maximum atom cloud density is attained.

The ash-atomize curves must be interpreted with caution. Graphite furnace atomization is prone to interferences due to chemical speciation of the analyte elements. For example the thermal pretreatment curves for an analyte present in real sample and in synthetic standard solution do not look necessarily the same. This is due to the fact that analyte may exist in many chemical forms in a real sample that can in principle be measured as such (waste water sample, for example). Thus a different behavior is observed when the thermal pretreatment temperature is raised. However, it is possible that due to different chemical forms present some portion of the analyte may escape already at low temperatures and the analyst may not observe this (even though

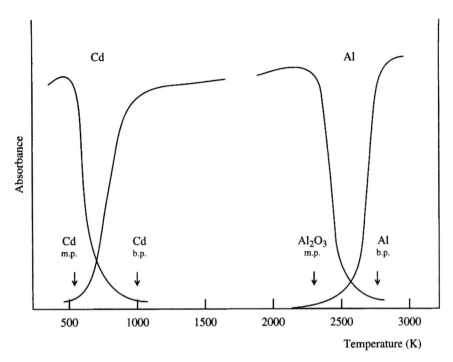

Figure 81 *Thermal pretreatment/atomization curve for cadmium and aluminium*

spiking experiments are made). Therefore it is sometimes necessary (and safe) to digest even a liquid sample to convert all chemical species to a one, known form.

It is possible to draw conclusions about the atomization mechanism from the thermal pretreatment/atomization curves, when melting point, boiling point, and decomposition point for the analyte and its compounds are entered on these plots. Figure 81 shows the thermal pretreatment/atomization curves for cadmium and aluminium. The initial pre-atomization losses and the appearance temperature are below the melting point of cadmium. This suggests that the atomization emanates from the metal. In the case of aluminium the atomization can be concluded to be a reduction of the oxide. The plot also suggests that the determination of aluminium in the cadmium matrix and the determination of cadmium in the aluminium matrix is possible without interference.

For many elements the following reduction mechanism has been postulated:

$$MO(s/l) + C(s) \rightarrow M(g) + CO(g) \tag{67}$$

However, from many observations and thermodynamic calculations this oxide reduction is not the only atomization mechanism coming into question. Three possible atomization mechanisms have been deduced from the activation energy values *(E_a)* obtained by a thermodynamic/ kinetic approach: (i) Thermal dissociation of the oxide; (ii) Thermal dissociation of the halide; (iii) Reduction of the oxide on hot graphite with subsequent volatilization of the metal. Later Sturgeon and Chakrabarti refined this procedure and proposed four different mechanisms:

(i) Reduction of the solid oxide on the graphite surface:

$$MO(s) + C(s) \rightarrow M(l) \nearrow 1/2\ M_2(g) \searrow M\ (g) \tag{68}$$

where M = Co, Cr, Cu, Fe, Mo, Ni, Pb, Sn, or V
(ii) Thermal decomposition of solid oxides:

$$MO(s) \rightarrow M(g) + 1/2\ O_2(g) \tag{69}$$

where M = Al, Cd, or Zn
(iii) Dissociation of oxide molecules in the vapour phase:

$$MO(s) \rightarrow MO(g) \rightarrow M(g) + 1/2 O_2(g) \tag{70}$$

where M = Cd, Mg, Mn, or Zn
(iv) Dissociation of halide molecules in the vapour phase:

$$MX_2(s) \rightarrow MX_2(l) \rightarrow MX(g) + X(g) \tag{71}$$
$$\downarrow$$
$$M(g) + X(g)$$

Frech and his colleagues studied the atomization processes in complex systems by means of high temperature equilibria calculations. These calculations showed that particularly O_2, H_2, and N_2 (from traces of water or nitric acid), and Cl_2 (from hydrochloric acid) were retained in the tube. For example, in the determination of lead even small amounts of chlorine in the sample matrix will cause the formation of volatile PbCl and $PbCl_2$ species. These species are driven from the furnace before the actual atomization temperature, so that substantial losses occur. It has been found that in the presence of H_2, greater concentrations of chlorine could be tolerated without the formation of noticeable amounts of lead chlorides. This is probably due to the removal of chlorine as hydrochloric acid.

In the determination of aluminium, CO, H_2, and Cl_2 interfere during the thermal pretreatment. $Al_2O_3(s)$ is stable up to 1800 K, in the presence of CO and H_2 losses occur at 1500 K. CO and H_2 are the main reaction products between graphite and water remaining in the tube. The state of the graphite is therefore important for the determination of aluminium, since no water remains on an intact surface layer of dense pyrolytic graphite. In addition O_2 and N_2 interfere during atomization, which means that these elements should be excluded for good results.

3.3.5.4 Platform Atomization. A platform is a thin, rectangular shaped piece of graphite (normally prepared from pyrolytic graphite), which is mounted inside the conventional graphite tube. The platform is positioned in the centre of the cuvette. The platform is heated by cuvette wall radiation since it fits into a slot in the wall of the graphite cuvette and therefore does not come into direct contact with the hot centre portion of the tube.

Since the platform is solely heated by the tube wall radiation, its temperature initially lags behind the tube wall temperature by several hundred degrees and takes several seconds longer to reach a constant temperature. The relationship between tube wall, gas phase, and L'vov platform temperatures for a longitudinally heated graphite tube is shown in Figure 63. The tube wall reaches the actual atomization temperature first, then the gas phase, and last the platform. At time t_3 the tube wall temperature is stable and the gas phase is hotter than the platform. Under these conditions the atomized analyte and matrix do not condense and therefore vapour phase interferences are minimized. Another advantage of platform atomization is a reduction in background signals due to greater thermal decomposition of molecules in the vapour phase compared to tube wall atomization. However, in the longitudinally heated furnaces there are no isothermal conditions in the gas phase because of the temperature gradients between platform and wall, and tube centre and tube ends. The latter of these gradients has been avoided by the transverse heated furnace (Figure 66). The platform provides a useful delay in atomization time, and in many cases improved sensitivity has been obtained (Figure 82). Peak shapes of arsenic for conventional graphite and pyrolytic graphite tubes and platform atomization are shown in Figure 83.

3.3.5.5 Probe Atomization. Attention has been paid to alternatives to platform atomization in graphite furnace AAS in order to get constant

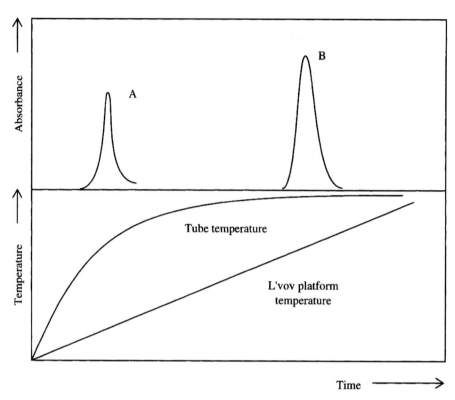

Figure 82   *Peak shapes for (A) tube wall and (B) platform atomization*

temperature conditions inside the graphite tube during the atomization stage. One such method is based on the introduction of the sample on a graphite probe which is introduced into the furnace atomizer only when stable atomization conditions have been established. The design and operation of the probe graphite furnace is presented in Section 3.3.1.

Probe atomization is advantageous for the determination of easy volatile elements. Many chemical interferences can be avoided by probe atomization. Figure 84 shows the interference of $MgCl_2$ on the determination of lead using tube wall, platform, and probe atomization. It is clearly demonstrated that probe atomization extends the useful interference free range in this particular example.

Another example of the usefulness of the probe atomization is the determination of lead in urine. Figure 85 shows the peak shapes obtained for an aqueous Pb standard and a urine sample containing the same amounts of lead. The peak shapes and appearance times are totally different for tube wall and platform atomization. Only in the case of probe atomization are the peak shapes, peak heights, and

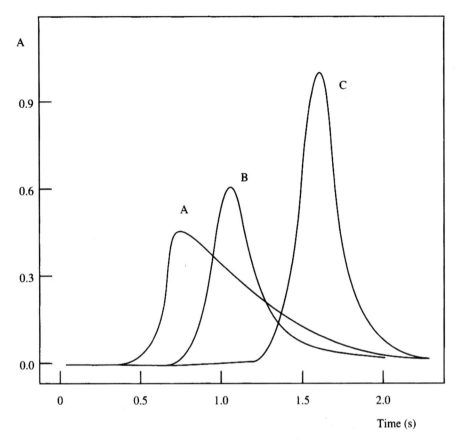

Figure 83 *Peak shapes of As (1 mg l⁻¹) obtained by (A) pyrolytic, (B) conventional, and (C) platform cuvettes*
(Unicam Analytical Systems Ltd.)

appearance times similar. This is an indication that both standards and samples are volatilized and atomized in a similar manner and the interferences have been eliminated.

3.3.6 Analysis of Solid Samples

In contrast to flame atomization, the graphite furnace technique offers relatively favourable conditions for the direct analysis of solid samples. Solid samples can be dispensed directly into the tube. Further, for solid samples thermal pretreatment stages and atomization stages are analogous for those of liquid samples. Several advantages of direct solid sampling can be obtained. The omission of a digestion stage can save time and simplify the entire analytical procedure. The sensitivity of atomic absorption can be fully exploited with solid samples, since the

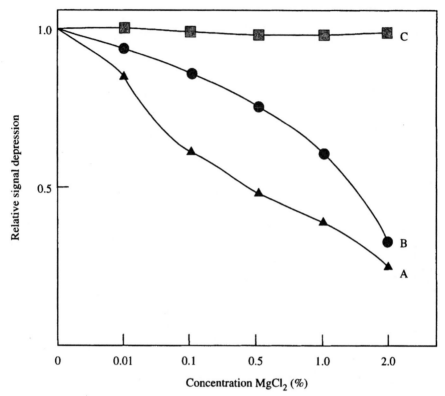

Figure 84 *The interference of MgCl₂ on the determination of Pb using (A) tube wall, (B) platform (with peak area measurements), and (C) probe atomization (with peak height measurements)*
(Unicam Analytical Systems Ltd.)

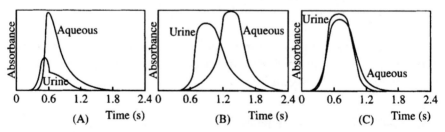

Figure 85 *Peak shapes for 50 µg l⁻¹ Pb in aqueous solution and in urine equivalent solution using (A) tube wall, (B) platform, and (C) probe atomization*
(Unicam Analytical Systems Ltd.)

concentration of the analyte is not diluted. Therefore further separation or enrichment steps are not needed. In trace elemental analysis every sample handling step, every reagent, and all laboratory ware can be sources of contamination or loss of the analyte. Thus, direct analysis of solids offers minimum systematic errors.

However, several disadvantages must be also taken into account: (i) The difficulty of accurately weighing and introducing a few milligrams of sample into the tube; (ii) With small weightings, homogeneity of the sample becomes a problem, leading to poor precision; (iii) Calibration is also difficult since solid samples cannot always simply be compared with aqueous reference solutions; (iv) Solid samples are mostly in the form of loose particles in the tube, which means that the heat transfer is poor because of the poor contact with the graphite surface; (v) In the case of organic samples, higher background attenuation than for dissolved samples can be expected.

Special cup tubes are designed for the direct analysis of solid samples (Figure 86). With this technique samples can be weighed directly into

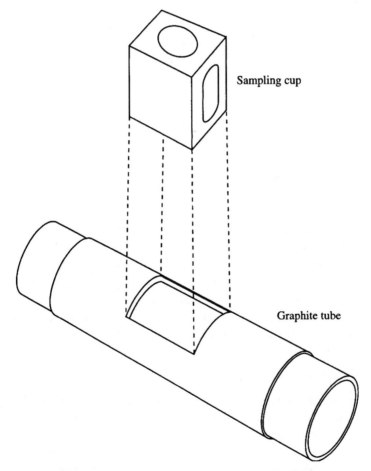

Sampling cup

Graphite tube

Figure 86 *Solid sampling cup for the cup-in-tube technique in GF-AAS*
(Adapted from: U. Völlkopf, Z. Grobenski, R. Tamm, and B. Welz, Analyst (London), 1985, **110**, 573)

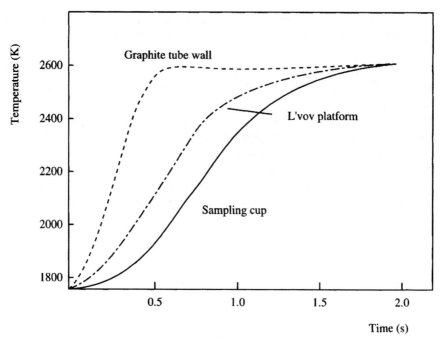

Figure 87   *Heating characteristics of (A) the inner wall of a pyrolytically coated graphite tube, (B) a L'vov platform, and (C) the inner surface of a solid sampling
cup*
(Adapted from: U. Völlkopf, Z. Grobenski, R. Tamm, and B. Welz,
Analyst (London), 1985, **110**, 573)

the cup. The cup acts as the L'vov platform and gives rise to the
temperature gradient between the tube wall and cup (Figure 87). When
cup tubes are used, aqueous standards are usually adequate for calibration since the cup significantly reduces background absorption and
other interferences during the atomization stage. Precision is about 10%
in most cases. Although the accuracy of the technique is not very good,
it is good enough for semiquantitative analyses.

3.3.7   Matrix Modification

Matrix modification (chemical modification) was proposed by Ediger in
1975 and it is a procedure for reducing or eliminating volatilization
and vapour-phase interferences. In this technique a reagent, usually
inorganic salt, is added (usually by an autosampler) to sample and
reference solutions in large excess. The reagent causes either (i) the
interfering concomitant to become more volatile, or (ii) the analyte to
be converted to a less volatile form. In both cases matrix modification

serves to make the volatilities of the analyte and concomitants sufficiently different to permit easier separation during thermal pretreatment. In the first case the concomitant can be removed before the atomization of the analyte. In the second case the analyte forms thermally more stable salts, oxides, or metal compounds with the matrix modifier. This enables the use of higher pretreatment and atomization temperatures. A large amount of matrix modifier may also act as a buffer that matches the chemical environment of aqueous standards and (undigested) samples more similar. Hence, for example problems related to a particular element existing in different chemical forms between aqueous standards and (undigested) samples is at least partly overcommed.

Chloride ion is an unwanted species in graphite furnace determinations. However, chloride matrix is often present in real samples (e.g. sea water, biological specimens), or chloride is introduced to samples during the sample preparation with hydrochloric acid (therefore nitric acid is preferred, whenever possible). For example, the addition of ammonium nitrate eliminates interferences due to sodium chloride according to the equation:

$$NH_4NO_3 \quad + \quad NaCl \quad \rightarrow \quad NaNO_3 \quad + \quad NH_4Cl \quad (72)$$

NH_4NO_3	$NaCl$	$NaNO_3$	NH_4Cl
decomposes	m.p. 1079 K	decomposes	subl. 618 K
at 483 K	b.p. 1691 K	at 653 K	b.p. 798 K

Sodium nitrate and ammonium chloride are formed, and these decompose or sublime at temperatures below 700 K. Excess of ammonium nitrate is also easily removed during the thermal treatment. Nitric acid can also be utilized as a matrix modifier to drive away chloride as hydrochloric acid during thermal pretreatment in graphite furnace program.

The large excess of the modifier reagent added has inevitably also some drawbacks in the practical determinations by GF-AAS. The modifier should not contain the element to be determined, or otherwise high blank values are obtained. The modifier solution used should preferably not contain elements that are later measured by the graphite furnace, since there is a risk for systematic errors due to memory effects. In that respect palladium, for example, is an ideal modifier element. Many modifier elements and their molecules (e.g. PO) may exhibit fine structured absorption spectra during atomization that in turn may cause over compensation errors, especially when continuum source background correction system is used. For instance, too high amounts of nickel modifier in the determination of selenium should be avoided.

The detailed function of a matrix modifier in real samples is very difficult to predict, especially when extremely low amounts of modifier elements and analytes are present in practical situations. However, some general comments can be made. First class of modifiers, e.g. magnesium and other alkaline earth elements, will form oxides in the graphite furnace during the thermal pretreatment. These oxides in turn form thermally stable mixed oxides with analyte elements and hence higher thermal pretreatment temperature can be used. Particularly magnesium (added as nitrate) is useful due to its high very high thermal stability. For the determination of aluminium, magnesium nitrate can be used as matrix modifier. During the thermal pretreatment stage, magnesium oxide is formed which encloses aluminium. The absorption signal of aluminium is obtained after evaporation of magnesium oxide (Figure 88).

The other class of modifiers consists of compounds that are reduced to metallic elements during thermal pretreatment. Noble metals (Rh,

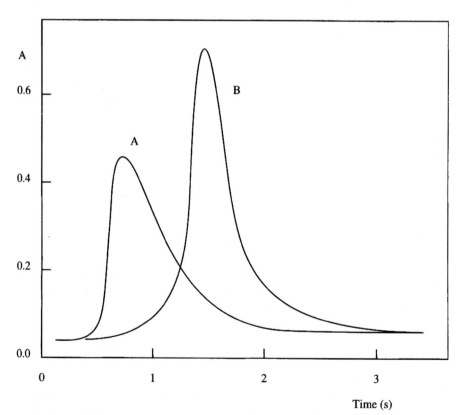

Figure 88   *The effect of Mg on the Al absorption signal. (A) Al signal without magnesium, and (B) magnesium added to the sample*

Pd, Ir, Pt) fall into this category. Occasionally a prereductant, such as ascorbic acid, is used in this connection to modify the properties of the reduced modifier. The reduced modifier (Pd and Rh, for example) evidently forms thermally stable intercalation compounds with graphite and the element determined is bind with the noble metal used. It is also possible that the analyte element form solid solutions (or compounds) with the reduced modifier element. In addition, the analyte elements are sometimes enclosed with the excess of the modifier used (in fact a reduced sensitivity results if too a large amount of modifier is used).

The function of the third class of modifiers may be understood in terms of the Lewis acid-base concept. Hence sulphate and phosphate, for example, act as hard Lewis bases and stabilize cations that are Lewis acids (e.g. Cd^{2+} and Pb^{2+}). Ammonium nitrate, ammonium hydrogen phosphate, and their mixtures can be used as matrix modifiers to reduce matrix effects in the determination of lead (recall also the acidic nature of these modifiers, cf. eq (72)).

Nickel is often used as matrix modifier in the determinations of arsenic and selenium (Ni^{2+} is a borderline Lewis acid). Nickel stabilizes arsenic up to about 1700 K, and selenium up to about 1500 K. This is presumably due to the formation of thermally stable nickel arsenide and nickel selenide. Nickel can also be used to stabilize bismuth, antimony, and tellurium. In addition, copper, silver, molybdenum, palladium, and platinum salts have been proposed as stabilizers for these elements. Palladium and platinum are the most suitable matrix modifiers for both inorganically and organically bound tellurium. Pretreatment temperatures up to 1320 K can be used. The effects of different matrix modifiers on the determination of antimony are shown in Figure 89. Without matrix modifiers the losses of antimony begin at 1000 K. Palladium, molybdenum, and nickel are the most suitable to stabilize antimony and with these modifiers pretreatment temperatures up to about 1600 K can be used. Recall, that the modifiers Cu, Ni, etc., may act as serious interferents in the liquid phase, when the elements mentioned above are determined using hydride generation method.

The addition of lanthanum for the determination of lead eliminates the interference of sulfate and increases the sensitivity. Phosphoric acid can be used to avoid losses of cadmium during the thermal pretreatment and to permit the use of higher atomization temperatures.

In addition to liquids, also gaseous and solid modifiers may be used in graphite furnace measurements. Oxygen may be added as an ashing aid during the thermal pretreatment. With an argon-hydrogen mixture a reducing environment is obtained, and also HCl can be

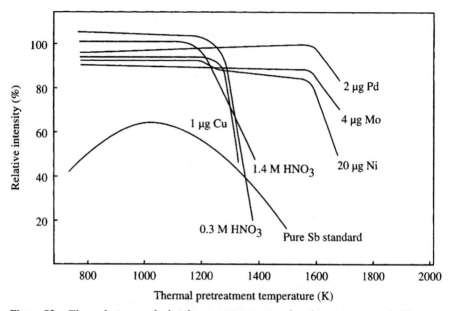

Figure 89   *The relative peak height sensitivities in the determination of Sb using various thermal pretreatment temperatures and matrix modifiers with the L'vov platform*
(Adapted from: H. Niskavaara, J. Virtasalo and L.H.J. Lajunen, Spectrochim. Acta, 1985, **40B**, 1219)

evaporated from the sample. The atomization of nonvolatile elements (e.g. uranium) can be assisted by using a fluorinating agent, such as CHF_3, as a modifier. Noble metal modifiers may be added as fine particles suspended to liquid (slurries).

During past few years the concept "permanent modifier" has been exploited. This means that the atomizer surface (graphite tube) is coated with a relatively stable layer of a modifier substance and thereafter even several hundreds of atomization cycles can be carried out (this holds at least for the elements that are atomized at low temperature, such as cadmium). Graphite furnace program is speeded up and also contamination problems are avoided. Several types of permanent modifiers are used. The noble metals that are volatilized at high temperatures (Ru, Rh, Ir, Pt) or elements forming stable carbides (e.g. W, Ta or Zr) are common examples. The graphite tubes coated with permanent modifiers can act also as efficient atom traps when used in connection with hydride generation method to further enhance the sensitivity of the atomic spectrometric determination of these elements.

Table 12 summarizes some matrix modifiers for the determination of various elements. Due to numerous sample matrices and elements

Table 12 *Matrix modifiers used for the determination of some elements*

Analyte element	Matrix modifier
Al	$Mg(NO_3)_2$
As	$Ni(NO_3)_2$
Be	$Mg(NO_3)_2$
Cd	$(NH_4)_2HPO_4$, NH_4F, EDTA, ascorbic acid, tartaric acid, HNO_3
Co	$Mg(NO_3)_2$
Cr	$Mg(NO_3)_2$
Ga	$HClO_4$
Hg	$(NH_4)_2S$, $K_2Cr_2O_7$, $KMnO_4$
Mn	$Mg(NO_3)_2$
Ni	$Mg(NO_3)_2$
P	$La(NO_3)_3$
Pb	$(NH_4)_2HPO_4$
Sb	$Pd(NO_3)_2$, $Mo(NO_3)_3$, $Ni(NO_3)_2$
Se	$Ni(NO_3)_2$
Sn	HNO_3
Ta	H_2SO_4
Te	$Ni(NO_3)_2$, $AgNO_3$
Tl	H_2SO_4

determined there is naturally a goal to simplify graphite furnace chemistry by the use of more universally applicable modifiers. One example is the use of Pd-Mg modifier solution that contains 10 µg of Pd and 15 µg of $Mg(NO_3)_2$.

3.3.8 Interferences in Graphite Furnace Atomization

In the case of flame atomization (Section 3.2.4) general interference effects involved in atomic absorption have been discussed. In this section only those interferences in graphite furnace atomic absorption occurring during the atomization process are described.

Interference effects in GF-AAS, may be categorized as being physical or chemical in the mechanism by which they are produced. Changes in sample introduction, background signals, and memory effects may give rise to physical interference effects. Chemical interferences are caused by the reaction of the analyte with a matrix component, tube material, or inert gas.

3.3.8.1 Physical Interferences. In GF-AAS, viscosity and surface tension of the sample solution have a less important part to play than in FAAS. They may affect the precision of micropipetting to some extent, but their main effect is in the degree to which the sample solution spreads inside the cuvette after the injection. This effect can be minimized by the ridged tube design and the use of peak area measurements.

Background absorption and scattering effects are the most serious interference sources in GF-AAS. In addition, continuum emission (particularly in the visible wavelength region) from the glowing graphite cuvette may distort the baseline and affect the photomultiplier response. Emission noise makes both accuracy and detection limit worse. This effect can be detected if the atomizer is run through its normal programme at the appropriate wavelength without any sample. Non-specific background absorption and scattering effects, and various background correction methods are discussed in detail in Section 3.4.

Memory effects are due to incomplete atomization or ineffective subsequent tube cleaning, and can result in an enhancement of the analyte response in later analyses. This problem is worst with elements forming very stable refractory oxides or carbides. This effect may be minimized by using very high atomization and tube cleaning temperatures, or longer times for these stages.

3.3.8.2 Chemical Interferences.   In GF-AAS chemical interferences may be classified into two groups: (i) Volatile compound formation, in which the analyte is wholly or partly lost from the graphite tube before the atomization temperature has been reached; (ii) Stable compound formation, where atomization is either completely or partly prevented or retarded by the formation of analyte compounds that remain unatomized at the normal atomization temperature.

The analyte may be lost during the thermal pretreatment phase for two reasons: (i) The analyte may be present in the sample as a compound which is appreciably volatile at the thermal pretreatment temperature used; (ii) The analyte may be converted into a volatile form by a matrix component or solvent. From the thermal pretreatment/ atomization curves (described in Section 3.3.5) it can immediately be seen, whether or not the thermal pretreatment temperature is too high. These plots also show the best thermal pretreatment and atomization temperatures for a given matrix.

Metal halides are in general much more volatile than the correspond-ing metal oxides. Therefore, oxyacids are preferred in sample prepara-tion to halogen acids for inorganic samples. Metal halides may be lost during the thermal pretreatment stage, or during the early part of the atomization stage and expelled from the cuvette before the atomization temperature is reached.

Tube-wall atomization should not be used in the determination of easily volatile elements. In the tube-wall atomization the analyte is vaporized to a gas-phase where the temperature is considerably lower than on the tube wall (Figure 63). This means that if an analyte is evaporated in molecular form it will probably atomize not at all. Otherwise, although free atoms are produced, they will reform molecules in the cool gas-phase. A very common example is the formation of various monohalides, since chlorine compounds are often present during the determination. The situation is especially complicated if an element is very easily volatilized (such as thallium), but it forms very stable monohalide (the dissociation energy of Tl-Cl is 373 kJ mol^{-1}). To obtain more interference free determination isothermal conditions should prevail during atomization. Therefore the use of platform (or probe) atomization is essential. The use of matrix modification will stabilize the analyte and delay its atomization unless higher gas-phase temperatures have been reached. Matrix modification is discussed in detail in Section 3.3.7. It this connection it can be mentioned that stable diatomic halides can be used in the determination of halogens by molecular absorption spectrometry (Section 3.5.6).

The response obtained for the analyte is reduced if it is present in the graphite tube as a thermally stable compound which is not easily dissociated in the environment of the tube at the atomization temperature. Most oxides are reduced at the atomization temperatures available in GF-AAS (Section 3.3.5). However, several elements can form carbides at temperatures lower than those required for gaseous atom formation. This would reduce considerably the response by slowing down the atom production rate, or preventing the atomization completely. Elements that form stable carbides are barium, vanadium, tungsten, molybdenum, and tantalum. Clean stage must always be used in the furnace programme when these metals are determined. The extent and ease of carbide formation depend on the state of the surface of the graphite tube. It should always be maintained in the same condition for reproducible results from such elements, and this is most easily achieved by using pyrolytic graphite tubes. Carbide formation may also be avoided by using tantalum or tungsten tubes or coating the graphite tube with tantalum.

If nitrogen is used as the inert shield gas, a few elements such as barium, molybdenum, and titanium may form stable nitrides.

3.4 BACKGROUND CORRECTION

Non-specific absorption is due to background absorption and scattering effects. This phenomena is more often encountered in electrothermal atomization than flame atomization. In the graphite furnace, light scatter is caused by mist and smokes due to the recombination of sample matrices at the cooler ends of the tube or from carbon particles shed from the tube walls. The light scattering increases rapidly with decreasing wavelength. Broad band molecular absorption is caused by molecules, radicals, or molecular ions formed or vaporized in the atomizer. The alkali and alkaline-earth halides, such as NaCl or $MgCl_2$, are particularly troublesome.

The amount of incident radiation absorbed or scattered must be measured and subtracted from the total measured absorbance in order to obtain the net absorbance of the analyte atoms only.

In all methods of background correction, two measurements are needed. The background correction requires that the spectrometer rapidly alternates between the total absorbance measurement and the background absorbance measurement, especially in electrothermal AAS where the absorption signal lasts only for a few seconds. A beam switching frequency of at least 150 Hz is required.

Background absorption can be compensated for or minimized by using sample-like standards or matrix modification, or by moving to an interference-free line if possible. However, the actual background correction methods are: (i) Two line method; (ii) Continuum source method; (iii) Smith-Hieftje method; (iv) Methods using the Zeeman effect.

3.4.1   Two Line Method

In the two line method the total absorbance (atomic absorption and background absorption) is measured at the resonance line of the analyte emitted by the hollow cathode lamp and the non-specific absorbance (background absorption) is measured at a nearby line where the analyte atoms do not absorb. Alternatively, another hollow cathode lamp of an element not present in the sample can be used. The line for the background absorbance measurement must be close to the analyte line. One of the problems is to find a convenient nearby line. The maximum permitted distance apart is 10 nm for most of the elements. To ensure there is accurate correction, the background must be constant within the range of these two lines. Another factor which must be considered is that for successive measurements the background signal are not necessarily constant, especially in the graphite furnace technique.

3.4.2   Continuum Source Method

The total absorbance is measured at the main resonance line of the element with a line-like radiation source. The non-specific absorbance is measured with a continuum radiation source and thus averaged over the spectral bandpass of the spectrometer (typically 0.2–0.7 nm). The width of the atomic absorption line is many times smaller than the monochromator bandpass so that the analyte absorption of the continuum radiation source is negligible. The continuum sources commonly used over the wavelength range 190–425 nm are the deuterium arc or hollow cathode lamp. A tungsten halogen source can be used over the wavelength range 300–900 nm. However, applications requiring background correction above 400 nm are very few.

By means of a rotating chopper with sector mirror, the radiation from the analyte line-like source and the radiation from a continuum source are passed alternately through the atomizer (Figure 90). Both radiation beams fall on the same detector after passing through the monochromator. The ratio of both radiation intensities is measured.

The principle function of a continuum source background corrector is depicted in Figure 91. The exit slit of the monochromator separates the resonance line of the analyte (half-width about 0.002 nm) from the emission spectrum of the line-like radiation source, and a band of radiation from the continuum spectrum of the deuterium lamp equivalent to the bandpass of the slit (usually 0.2 to 0.7 nm). The intensity of the hollow cathode lamp (I_{hcl}) is equalized to the intensity of the deuterium lamp (I_{dl}) before the determination. When the ratio $I_{hcl}/I_{dl} = 1$ no reading shows on the display. When a solution containing analyte element is introduced into the atomizer both intensities (I_{hcl} and I_{dl}) are attenuated due to atomic absorption. Since half-width of the resonance

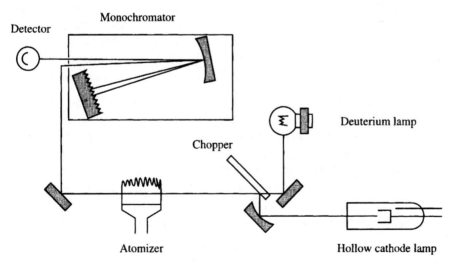

Figure 90 *Schematic configuration of an AA spectrometer with a continuum source background corrector*

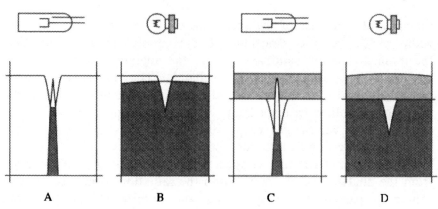

Figure 91 *Mode of operation of a continuum source background corrector. A and B: Atomic absorption only; C and D: Atomic absorption with background absorption*

line is only about 0.002 nm while the continuum has a width of at least 0.2 nm, even for 100% absorption of the radiation from the hollow cathode lamp, a maximum of 1% of the continuum would be absorbed. Thus, in normal measurements, atomic absorption of the continuum is less than 1% and can be neglected. Consequently I_{dl} is unattenuated when reaching the detector, and I_{hcl} is attenuated proportionally to the concentration of the analyte in the sample. The ratio I_{dl}/I_{hcl} is now greater than 1, and a reading is obtained.

If non-specific attenuation occurs, both radiation beams will be attenuated to the same extent provided that the background attenuation is a broad band phenomenon. Since background absorption attenuate I_{dl} and I_{hcl} to equal degrees, the ratio I_{dl}/I_{hcl} remains unity. Thus, no reading is obtained and the background is corrected. If analyte specific absorption takes place in addition to the background attenuation, I_{hcl} will be further attenuated, leading to a normal absorbance reading.

3.4.2.1 Continuum Radiation Sources. Continuum radiation sources used for background correction are deuterium arc lamps, deuterium hollow cathode lamps, xenon mercury arc lamps, and tungsten halogen lamps. These operate over the wavelength range 190–425 nm, except tungsten halogen lamps which operate between 300 and 900 nm. Deuterium hollow cathode lamps are simpler than deuterium arc lamps, but their intensity is much lower (Figure 92). Xenon mercury arc lamps are stable and their radiation intensity is high over wide wavelength ranges.

3.4.2.2 Faults in Background Correction. Background correction with a continuum source has some disadvantages and limitations. The noise is increased by a factor of two or three due to the second radiation

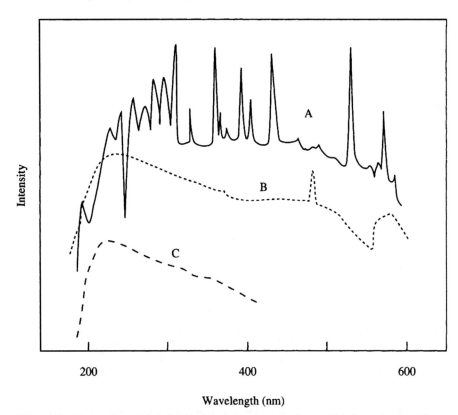

Wavelength (nm)

Figure 92 *Emission spectra of (A) xenon-mercury arc lamp, (B) deuterium arc lamp, and (C) deuterium hollow cathode lamp*
(Adapted from: M. S. Epstein and T. C. Rains, Anal. Chem., 1976, **48**, 528)

source and the altered signal handling. Strong background attenuation increases the noise further, which means the background attenuation should be kept below 0.5 absorbance units.

Spectral interferences may be overcompensated (or undercompensated) using continuum background correction. If the sample contains an element which has an absorption line so close to the analyte resonance line that it will pass through the monochromator bandpass, the measured background absorption is too high. As a result too low an analyte concentration is obtained. For example, antimony gives rise to overcompensation in the determination of lead when using a deuterium hollow cathode lamp for background correction. The degree of overcompensation depends on the slit width. The background absorption signal decreases with decreasing slit width, and reaches a constant value at 0.4 nm (Figure 93). Then there is no more overcompensation.

The overcompensation errors might be serious in GF-AAS determinations where the concentration of analyte is very low, but the

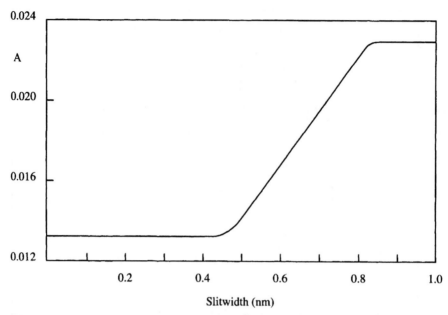

Figure 93 *The effect of slit width on the background absorption at the wavelength of the Pb resonance line 217.0 nm, when Sb solution (100 mg l⁻¹) is introduced into the air-acetylene flame using the hydrogen hollow cathode lamp correction* (Adapted from: F. Vaijda, Anal. Chim. Acta, 1981, **128**, 31)

concentrations of matrix elements are high. In such cases weak secondary absorption lines of the matrix elements may result in serious overcompensation errors. A well known interferent is iron that is present at high concentrations for example in many biological samples (Figure 94).

Continuum source background correctors are incapable of correct background attenuation of electronic excitation spectra, which consist of many narrow lines. These spectra originate from electronic transitions within the molecule. The structures of the bands are due to transitions from the rotational and vibrational levels of one electronic state to those of another electronic state. In this case, the background correction depends on the degree of overlap between the spectral line of the analyte and the individual molecular rotational lines. This demands much higher resolution power than the capabilities of the monochromators used in AAS. For example, the resonance line of gold at 267.6 nm lies exactly between two rotational lines of indium chloride, so that the actual background attenuation is relatively low. If a continuum source is now employed for the background correction, overcompensation will occur since the corrector determines the mean absorbance over the observed spectral range.

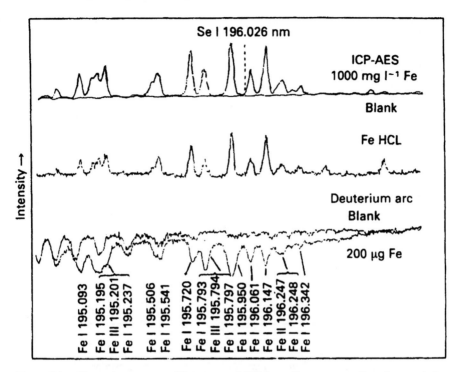

Figure 94 *Absorption spectra of iron around 196.0 nm Se resonance line. Iron emission*
spectra of hollow cathode lamp and ICP source are also shown
(Adapted from: I. Martinsen, B. Radziuk and Y. Thomassen, J. Anal. At.
Spectrom., 1988, **3**, 1013)

The background correction with a continuum radiation source is
an extremely useful device and leads frequently to the correct results.
However, for samples of unknown composition, a check should be
made to see whether correction is complete or not.

3.4.3 Smith-Hieftje Method

In the Smith-Hieftje method the emission line profile of the radiation
source is modified to measure the background. When extremely high
currents are applied to a hollow cathode lamp, the emission line is considerably broadened with a dip appearing in the centre of the emission
profile due to self-absorption (Figure 95). At low current the narrow
emission line from the hollow cathode lamp is used to measure the
absorbance of both analyte and background. At high lamp current
the measured absorbance is essentially that due to the background.
Thus, the analyte absorbance is simply the difference of these two
absorbances. However, there will be some absorption by the analyte at

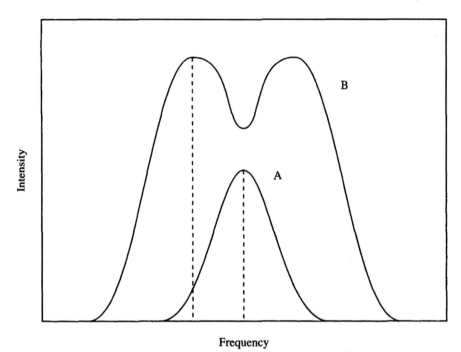

Intensity

Frequency

Figure 95   *The principle of the Smith-Hieftje background correction. The emission profile emitted by a hollow cathode lamp by using (A) low current, and (B) high current*
(Adapted from: S. B. Smith, Jr. and C. M. Hieftje, Appl. Spectrosc., 1983, 37, 419)

high lamp current, leading to a loss of sensitivity of a factor of perhaps two or three, depending on the element.

The advantages of the Smith-Hieftje method are similar to those of the Zeeman background correction (however, the Zeeman system is much more expensive): (i) The use of a single radiation source for any wavelength; (ii) Alignment of the source is not a problem like in continuum source method; (iii) The ability to correct at a wavelength very close to the analyte absorption line; (iv) Highly structured backgrounds can be corrected.

The disadvantages of this method are: (i) The electronics cannot be economically incorporated into an existing spectrometer due to the extensive changes which would be required in the electronics; (ii) The lifetime of the hollow cathode lamps for the more volatile elements is inevitably shortened; (iii) 'Roll-over' phenomenon, analogous to that observed in Zeeman background correction (Section 3.4.4), exists at higher concentrations.

The roll-over phenomenon is due to the fact that the corrected signal is obtained as a difference of two absorption signals. Respectively, a calibration graph is obtained as a difference of two non-linear graphs. These graphs show different slopes at various concentration levels, depending on the stray light present during the both absorption measurements made. Hence, the same kind of double peak in GF-AAS measurement is observed than in measurements utilizing Zeeman effect correction (Figure 103).

3.4.4 Methods Achieving Zeeman Effect

Zeeman background correction can be put into operation in several different ways and the method employed can have a direct effect on the performance of the instrument. The total absorbance (specific and non-specific) and the background absorbance are measured by the same line-like radiation source. The use of Zeeman effect allows the atomic absorption and background absorbances to be measured at the same wavelength or the background signal very close to the atomic absorption signal. The magnet can be positioned either (i) around the radiation source (Direct Zeeman AAS) or (ii) around the atomizer (Inverse Zeeman AAS).

3.4.4.1 Direct Zeeman AAS. In this technique the magnet is positioned around the lamp so that the emission profile from the line-like radiation source is split into the π- and σ-components as shown in Figure 96. A rotating polarizer is required to distinguish and isolate the components parallel (π-component) and perpendicular (σ-components) to the magnetic field. The π-emission signal is used to measure the total absorbance (specific and non-specific absorption) at the analyte resonance line. The σ-signals are at slightly displaced wavelengths and the resulting absorbance measured represents background absorbance only. The strength of the magnetic field applied must be high enough so that the σ-components can be isolated from the absorption line. Normally, field strength of about 10 kG is adequate, and both σ-components are then displaced about 0.01 nm.

In the case of normal Zeeman effect, the position of the π-component is unchanged and independent of the magnetic field strength. On the contrary, in the case of anomalous Zeeman effect the strength of the field has an influence also on the π-component, and the absorbance will decrease with increasing field strength. Figure 97 shows the absorption signals for cadmium and silver at the wavelength 228.8 nm and 328.1 nm, respectively, as a function of the magnetic field strength. For

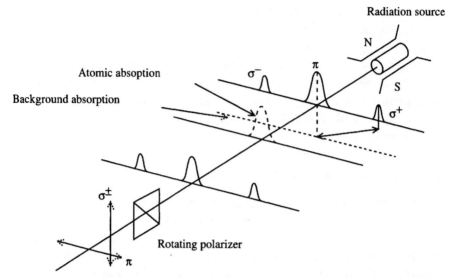

Figure 96 *The direct Zeeman AAS technique in which the magnet is positioned around the lamp*
(Adapted from: T. J. Hadeishi and R. D. McLaughin, Science, 1971, **174**, 404)

cadmium (normal Zeeman effect) the signal of the π-component is constant over the range of the field strength, whereas for silver the signal of the π-component decreases with increasing field strength. In both cases the signals of the σ-components decrease with increasing field strength.

Advantages of this technique are: (i) Only one light source is used; (ii) The technique is not limited to radiation sources operating in the UV region.

Disadvantages of direct Zeeman AAS are: (i) Non-absorbed lines from the radiation source must be absent as in ordinary atomic absorption spectrometry; (ii) The background is not measured directly at the resonance wavelength so that poor correction will occur, especially when the background is structured; (iii) It is difficult to initiate the hollow cathode lamp discharge in a strong magnetic field, so special radiation sources are needed (iv) The stability and intensity of the radiation are affected by the magnetic field and the polarizer required.

3.4.4.2 Inverse Zeeman AAS.   In the inverse Zeeman AAS methods the magnet is positioned around the atomizer (either burner or graphite furnace). Thus, in this case the absorption profile of the analyte atoms is split into the π- and σ-components. Depending on the direction of the magnetic field with respect to the radiation beam, inverse Zeeman

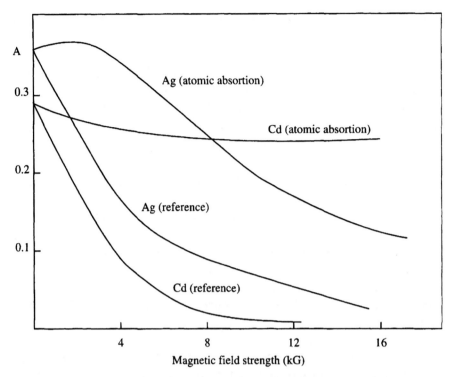

Figure 97 *Comparison of 'sample' and 'reference' absorption signals with increasing magnetic field for lines showing normal (Cd, 228.8 nm, $^1P_1-^1S_0$) and anomalous (Ag, 328.1 nm, $^2P_{3/2}-^2S_{1/2}$) Zeeman splitting* (Adapted from: R. Stephens and D. E. Ryan, Talanta, 1975, **22**, 659)

AAS is classified into two groups: transverse and longitudinal Zeeman AAS. In the case of transverse Zeeman effect a constant or rotating polarizer is also required to separate the radiation beam from the hollow cathode lamp into two components, one polarized parallel to the magnetic field and the other perpendicular to the field. In longitudinal Zeeman AAS no polarizer is needed.

In the transverse AC Zeeman, the polarizer is positioned either between the radiation source and the atomizer or between the atomizer and monochromator. The π-component of the absorption profile interacts with the radiation polarized parallel to the field and produces the total absorbance signal (specific and non-specific absorption). The radiation polarized perpendicular to the field interacts only with the non-specific absorption and therefore this can be used to measure the background absorption.

The most significant advantage of inverse Zeeman AAS is that the background is always measured directly at the same wavelength as the atomic absorption of the analyte. This permits accurate correction even when the background is highly structured, but not when non-absorbed lines are present.

In inverse Zeeman AAS there are some practical constraints on the atomizer due to the need to have the magnet pole-pieces as close together as possible in order to keep the field strength (and the power supply required) as small as can be achieved. Hence, in the case of the flame atomizers, water cooling is necessary to prevent overheating of the magnet pole-pieces.

The magnetic field can be produced by a constant or changeable DC magnet, or by an AC magnet. The field can be parallel to the radiation beam or perpendicular to it. The design of an instrument in which the field is parallel with the radiation beam is technically difficult, and therefore in most commercial instruments the field is perpendicular to the radiation beam. The transverse Zeeman AAS is classified on the basis of the magnet type, *i.e.* (i) Direct current Zeeman AAS (DC Zeeman AAS), and (ii) Alternating current Zeeman AAS (AC Zeeman AAS).

(i) *Transverse DC Zeeman AAS:* The general construction of an inverse Zeeman atomic absorption spectrometer equipped with a DC magnet is presented in Figure 98. In this case, magnets which give constant magnetic fields are generally employed and hence the Zeeman effect is permanently applied. The rotating polarizer divides the radiation beam from the hollow cathode lamp into the two linearly polarized components. One of the components oscillates parallel to the magnetic field ($P_{\parallel}$), and the other one perpendicular to it ($P_{\perp}$). These components have the same wavelength. The $P_{\parallel}$-emission profile interacts with the resonance line of the analyte element (π-component) and background absorption, whereas the $P_{\perp}$-emission profile interacts only with the background emission since the σ-components (oscillating perpendicular to the field) are shifted due to the magnetic field. The absorption line profile is broadened as a result of the Doppler and Lorentz broadening effects, but the σ-components can be isolated from the absorption line by a suitable field strength.

A constant field may be simpler and cheaper to obtain than an alternating field, but there are several problems encountered with this system:

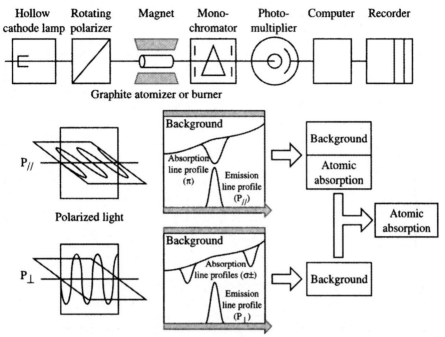

Figure 98 *Schematic configuration and operation of a DC Zeeman AAS instrument in which the magnet is positioned around the sample*
(Adapted from: H. Koizumi and K. Yasuda, Spectrochim. Acta, 1976, **31B**, 237)

(1) The rotating polarizer is required to separate the if π- and σ-components which bring two problems: the polarizer must rotate at quite high speeds in order to achieve fast modulation capable of compensating backgrounds where fast transient electrothermal atomizer peaks are involved, and there is a large loss of light in the UV region because of the polarizer material itself.

(2) Intensities of the radiation beams are dependent on the field strength, absorption line, and isotope constitution of the element. Sensitivity for elements with anomalous Zeeman patterns (Ag) depends on the magnetic field strength (Figure 99). For Pb (with the normal Zeeman effect) the $P_\parallel$-absorption is only a little dependent on the magnetic field strength. A small decrease in absorbance is due to the Pb-isotope (21.1%) with a nuclear spin of 1/2. The optimum field strength for the elements with a normal Zeeman pattern is about 9–10 kG, whereas that for the elements with an anomalous Zeeman pattern varies between 3 and 16 kG (Table 13). With permanent magnets it is impossible to have optimum field strength for all elements. However, with DC electromagnets the field strength may be optimized by altering

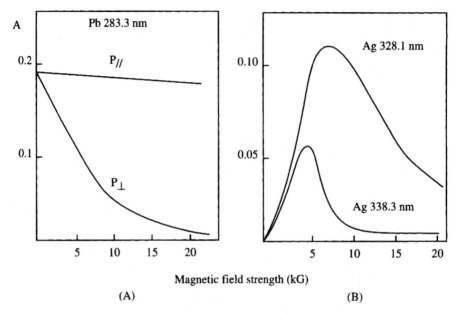

Figure 99 *(A) Absorbances of the 'sample' ($P_{\parallel}$) and 'reference' ($P_{\perp}$) signals of lead as a function of increasing field strength, and (B) differences of the 'sample' ($P_{\parallel}$) and 'reference' ($P_{\perp}$) signals of silver with increasing field strength* (Adapted from: H. Koizumi and K. Yasuda, Spectrochim. Acta, 1976, **31B**, 237)

Table 13 *Optimum magnetic field strengths for some elements for indirect DC Zeeman-AAS*

Element	Zeeman effect*	Optimum magnetic field strength (kG)
Ag	A	5
Al	A	10
As	A	5
Ba	N	9
Cd	N	9
Cu	A	13
Cr	A	4
Ir	A	16
Mo	A	3
Pb	N	8
Zn	N	10

*A = anomalous, N = normal.

the power applied (and hence the field). Even when the field has been optimized, the sensitivity for elements with an anomalous Zeeman pattern is always lower than for elements with a normal Zeeman pattern when the magnetic field is permanently applied.

(3) The linear dynamic range for the elements with the anomalous Zeeman pattern depends on the position of the magnet and the field strength applied.

(ii) Transverse AC Zeeman AAS: The inverse AC Zeeman background correction methods may be distinguished by the alignment of the magnetic field with respect to the optical axis. In transverse AC Zeeman effect the magnetic field is perpendicular to the optical axis, and in the longitudinal direction the magnetic field is parallel to the optical axis.

In transverse AC Zeeman AAS the magnet is powered by an alternating current so that the field is successively switched on and off (Figure 100). The frequency of the modulated magnet is either 50 or 60 Hz depending on the modulation frequency of the hollow cathode lamp (100 or 120 Hz, respectively). When the magnetic field is switched on, the σ-components are displaced provided the magnetic field strength is high enough. Usually, a magnetic field strength of 8 kG is adequate. The principle of the transverse AC Zeeman effect background correction is shown in Figure 101. The total absorbance (atomic absorption and background absorption) is measured when the magnet field is off, like in conventional AAS. When the field is switched on, the absorption profile of the analyte is split into the π- and σ-components. A static polarizer positioned between the atomizer and monochromator rejects the π-component so that only the background is measured when the field is on. The actual atomic absorption signal of the analyte atoms is again obtained by the difference between these two absorbance measurements.

The sensitivity obtained by transverse AC Zeeman AAS for the elements with the normal Zeeman pattern is good and often nearly the same as for the conventional AAS. In addition, for many elements with the anomalous Zeeman effect such as Ni, Mn, Sb, Tl, and Ag, good sensitivity is obtained. In general, the sensitivities are better with an AC magnet than with the permanent DC magnets, also the linear dynamic range for the AC magnet is better being about the same as for conventional AAS. However, the polarizer in this system also diminishes the observed signal since the polarizer material is absorbing radiation in the UV region. The π-component is linearly polarized parallel to the magnetic field and has an intensity of 50% compared to the incident radiation. The σ-components are linearly polarized perpendicular to the magnetic field and have a total intensity of 50% compared to the incident radiation.

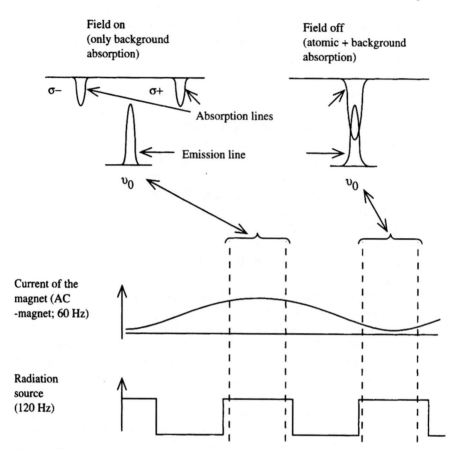

Figure 100 *Background correction with the transverse AC Zeeman AAS technique*
(Zeeman 5000, Perkin Elmer Corp.)

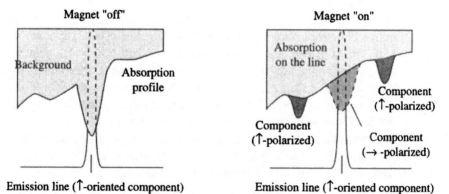

Figure 101 *Principle of the transverse AC Zeeman effect background correction*
(Perkin Elmer Corp.)

Benefits for the transverse AC Zeeman effect background correction are: *(a)* The background correction is performed exactly at the same wavelength as the atomic absorption measurement; *(b)* High nonspecific absorbances (up to about 2) can be corrected; *(c)* When the magnetic field is switched on, only non-specific absorption is measured; *(d)* When the magnetic field if switched off, the total absorbance (specific and non-specific) is measured as in conventional AAS; *(e)* Sensitivities and linear dynamic ranges are comparable to those for conventional AAS; *(f)* Single-beam instrument optics and a static polarizer are required; *(g)* In practice, detection limits are smaller than in conventional AAS.

There are practically no differences in sensitivities obtained by a permanent DC magnet or by an alterable DC magnet for elements with the normal Zeeman effect. The sensitivity increases with increasing field strength and reaches the maximum value when the field strength is about 10 kG. For the elements with the anomalous Zeeman effect the sensitivity first increases, reaches its maximum value and then decreases with the increasing DC magnet field. The maximum sensitivity for these cases is always lower than that for the corresponding elements by conventional AAS. The optimum field strength varies from one element to another, which means that with a permanent DC magnet optimum conditions cannot be obtained for all the elements. The sensitivities for most of the elements when using the transverse AC Zeeman background correction is about 70–90% of the corresponding sensitivities by conventional AAS (Table 14).

In conventional atomic absorption measurements the measured signal depends only on one absorption coefficient, whereas the Zeeman AAS signal depends on two absorption coefficients (Figure 102). Hence, the calibration graph obtained is the difference between these two nonlinear curves, and it may have a strong curvature at the end of the graph. For the spectral lines with the normal Zeeman effect the calibration graph is essentially similar to that of the conventional AAS, whereas for the spectral lines with the anomalous Zeeman effect, the shape of the calibration graph depends on the position of the magnet, the type of the magnet (DC, AC), and the magnetic field strength. In these cases the calibration graph is strongly curved, and the same absorbance value is obtained at two different analyte concentrations. This phenomenon is called 'roll-over', and the maximum point on the curve is the 'roll-over point'. At analyte concentrations higher than the

Table 14   *Comparison of sensitivities of DC-ZAAS and AC-ZAAS with respect conventional AAS[a]*
(Source: W. Bohler, M. M. Beaty and W. B. Barnett, At. Spectrosc., 1981, **2**, 73)

Analyte	Wavelength (nm)	Zeeman effect[b]	DC-ZAAS constant field[c]	DC-ZAAS variable field[d]	AC-ZAAS
Ag	328.1	A	0.28	0.61	0.91
Al	309.3	A	0.97	0.97	0.90
As	193.7	A	0.43	0.66	0.89
Au	242.8	A	0.56	0.63	0.84
B	249.7	A	0.62	0.62	0.61
Ba	553.6	N	0.97	0.97	0.96
Be	234.9	N	0.62	0.62	0.51
Bi	223.1	A	0.71	0.71	0.63
Ca	422.7	N	0.97	0.97	0.95
Cd	228.8	N	0.94	0.94	0.98
Co	240.7	A	0.87	0.87	0.85
Cr	357.9	A	0.42	0.68	0.88
Cu	324.8	A	0.49	0.49	0.53
Fe	248.3	A	0.73	0.76	0.92
Ge	287.4	N	0.85	0.85	0.90
Hg	253.7	N	0.71	0.71	0.61
In	325.6	A	0.82	0.82	0.91
Ir	264.0	—	0.91	0.91	0.96
K	766.5	A	0.28	0.65	0.61
Li	670.7	N	0.42	0.42	0.85
Mg	285.2	A	0.81	0.81	0.91
Mn	279.5	A	0.61	0.65	0.95
Mo	313.3	A	0.28	0.61	0.95
Na	589.5	A	0.35	0.65	0.95
Ni	232.0	A	0.88	0.88	0.91
P	213.6	A	0.78	0.78	0.70
Pb	283.3	N	0.86	0.86	0.83
Pd	247.6	N	0.91	0.91	0.91
Pt	265.9	—	0.80	0.80	0.80
Ru	349.9	A	0.86	0.86	0.90
Sb	217.6	A	0.68	0.81	0.95
Se	196.0	A	0.45	0.51	0.88
Si	251.6	N	0.94	0.94	0.98
Sn	286.3	N	0.97	0.97	0.94
Sr	460.7	N	0.75	0.75	0.98
Te	214.3	A	0.79	0.82	0.93
Ti	365.3	A	0.94	0.97	0.96
Tl	276.8	A	0.46	0.51	0.66
V	318.4	A	0.70	0.79	0.77
Zn	213.8	N	0.90	0.90	0.88

[a]Sensitivity for the conventional AAS = 1, [b]A = anomalous and N = normal, [c]10 kG, [d]maximum strength of the field 10 kG.

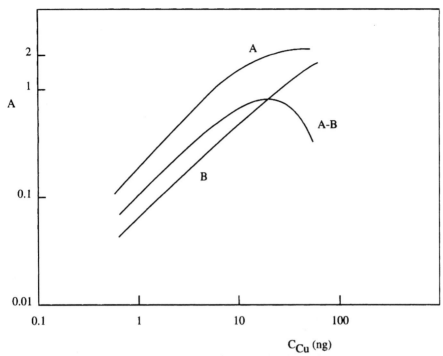

Figure 102 *Calibration graphs for copper (λ = 324.8 nm) obtained by transverse AC Zeeman AAS. A: magnet 'off'; B: magnet 'on'; A-B: ZAAS signal (difference between A and B)*
(Adapted from: F. J. Fernandez, W. Bohler, M. M. Beaty and W. B. Barnett, At. Spectrosc., 1981, **2**, 73)

roll-over point, two absorption peaks are observed (Figure 103). It is possible to monitor the appearance of the double peaks with the micro-computer of the instrument.

(iii) Longitudinal AC Zeeman AAS: The Zeeman 4100ZL by Perkin Elmer is the first atomic absorption spectrometer using the longitudinal Zeeman effect. In this technique the π-component is totally missing and the σ-components are circularly polarized and have an intensity equal to that of the incident radiation. With this arrangement the absorption line splitting is such that no absorption except the real background absorption is measured on the original resonance line of the analyte when the magnetic field is on. The great advantage this provides is that no polarizer is required and the total light energy can be used for measurements. The principle of the longitudinal AC Zeeman background correction is displayed in Figure 104. Signal processing in the Zeeman 4100ZL is performed such that when the field is switched on two measurements (background

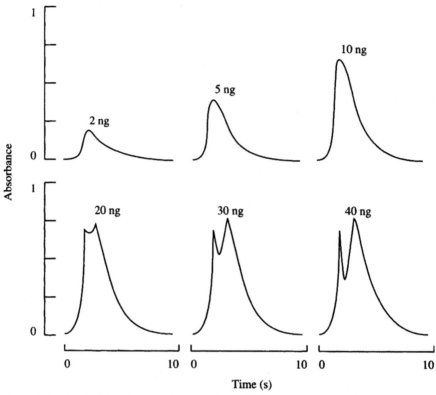

Figure 103   *Peak shapes for copper at 324.8 nm showing the 'roll-over' effect*
(Adapted from: F. J. Fernandez, W. Bohler, M. M. Beaty, and W. B.
Barnett, At. Spectrosc., 1981, **2**, 73)

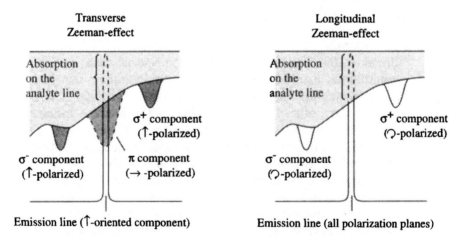

Figure 104   *Principle of the longitudinal AC Zeeman AAS background correction*
(Perkin Elmer Model 4100 ZL Atomic Absorption Spectrometer)

absorption) are taken before and after the measurement when the field is switched off (atomic absorption and background absorption). The average of the field-on measurements is subtracted from the respective field-off measurement. This procedure guarantees accurate results even when the background is changing rapidly.

The background correction by the longitudinal Zeeman effect is ideal since it dispenses with the need for polarizers to separate out of the components. Advantages are: (i) Accurate correction of very high background; (ii) Accurate correction of rapidly changing background; (iii) True double beam operation by alternately measuring the total absorption and the background absorption with the same identical optical path; (v) Maximum light energy throughput and stable baseline; (vi) Good signal-to-noise ratio and very low detection limits.

3.5 SPECIAL METHODS

The determination of trace elements such as Pb, Cd, Zn, Cu, As, Se, and Hg in environmental or clinical samples is often required at sub p.p.m. levels, the extreme limits of sensitivity for conventional AAS. At these low concentrations these elements can be readily determined by graphite furnace AAS. However, GF-AAS is relatively expensive, time-consuming, and in some cases too sensitive (for example, Zn in serum) a method. Typically, the measurement takes 2 to 3 minutes by GF-AAS, while by FAAS it takes 10 seconds or less.

A number of methods have been described for improving the sensitivity of conventional FAAS in order to allow the analysis without resorting to more expensive techniques. Best known of these techniques are hydride generation, cold vapour, semi-flame (Delves cup, tantalum boat), and slotted tube atom trap (STAT) methods.

Some elements cannot be determined by conventional AAS. The most suitable lines of non-metals (halogens, S, P, C, O, N) are located in the vacuum UV region. Some elements are difficult to atomize (Th, Ce, U, La, Nb, Zr). The detection limits for some elements (As, Ge, Ti, V, Se, Te) are too high to allow determinations at trace concentrations. Several indirect methods have been developed to determine these elements and also organic compounds.

3.5.1 Slotted Tube Atom Trap

The use of a slotted quartz tube to increase the sensitivity of the flame AAS technique was first described by Watling in 1977. A double slotted tube was supported above an air-acetylene flame from a conventional burner with one of the slots aligned directly above the flame. The tube was also aligned with the optical path of the spectrometer so that there was minimum attenuation of the radiation from the light source on passing through the flame. The slots were machined laterally so that they were parallel to each other at 120°. Sensitivity enhancements of three or five times for many easy volatile elements were obtained. The enhancement effect is due to the prolonged residence time of the analyte atoms in the tube. Watling studied both single and double slotted tubes. There was practically no difference in sensitivity, but the single slotted tubes gave greater memory effects, and directed the flame out of the tube ends onto the optical windows. The flame exits through the extra slot in the double slotted tubes rather than the tube ends. Performance figures obtained with the STAT technique for some volatile elements are given in Table 15.

3.5.2 Hydride Generation Methods

The separation of the analyte from the matrix is a major advantage of the hydride generation method. The analyte passes into the atomizer as a gaseous hydride, while concomitants normally remain in the reaction vessel. Spectral interferences can virtually be excluded since a relatively small number of components are present in the atomizer.

Initially in the hydride generation methods, atomization was performed mainly in argon/hydrogen diffusion flames, and later these were replaced by graphite furnaces or heated quartz tubes. Nowadays, commercial hydride generation systems use almost exclusively either

Table 15 *Sensitivities and detection limits (mg l^{-1}) by conventional and STAT (Slotted tube atom trap) FAAS methods for some volatile elements* (Source: A. A. Brown and S. F. N. Morton, Laboratory Science and Technology, September, 1985)

Analyte	Flame	STAT		Conventional FAAS	
		Sensitivity	DL	Sensitivity	DL
Pb	Air-C_2H_2	0.03	0.0035	0.10	0.005
As	Ar-H_2	0.06	0.008	0.30	0.09
Se	Ar-H_2	0.08	0.015	0.26	0.10
Cd	Air-C_2H_2	0.004	0.0005	0.01	0.001
Cu	Air-C_2H_2	0.015	0.001	0.035	0.002
Sn	Air-H_2	0.10	0.015	0.35	0.06
Zn	Air-C_2H_2	0.004	0.0004	0.010	0.0008

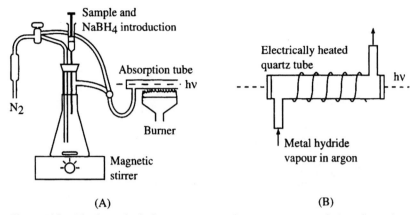

Figure 105 *The basic hydride generation technique systems with (A) flame heated and (B) electrically heated quartz tubes for atomization*

electrically heated or flame heated quartz tubes for atomization. The basic systems are represented in principle in Figure 105. In each case a standard atomic absorption spectrometer is used. A quartz absorption tube is either positioned above the burner (Figure 105A) or the burner assembly of the spectrometer is replaced with a quartz tube which is heated electrically by a coiled wire (Figure 105B). The hydride vapour is atomized in the quartz tube and the resulting atoms are detected by the radiation beam from the hollow cathode lamp (or EDL) which passes down the axis of the tube. Non-specific background absorption is almost unknown with this technique when heated quartz tubes are used.

Another, earlier a widely used hydride generation system is represented in Figure 106. A small oxygen-hydrogen flame burns inside an unheated quartz tube. The hydride vapour is flushed by a stream of hydrogen into the tube where it is atomized in the hot oxygen-hydrogen flame. Excess hydrogen is burnt at the ends of the tube. The flow rate of hydrogen is about $1.5 \, l \, min^{-1}$ and that of oxygen about $6 \, ml \, min^{-1}$. In this technique oxygen-hydrogen and air-hydrogen flames cause background absorption which must be corrected.

In heated quartz tubes no interferences due to background attenuation occur provided that hydrogen is not allowed to ignite at the open tube ends. For atomization in flames the background correction is recommended since variations in the transparency of the flame may occur when the hydride and the hydrogen generated enter the flame. When atomization takes place in an oxygen-hydrogen flame inside an unheated quartz tube with hydrogen-air flames at the ends of the tube, background correction is needed.

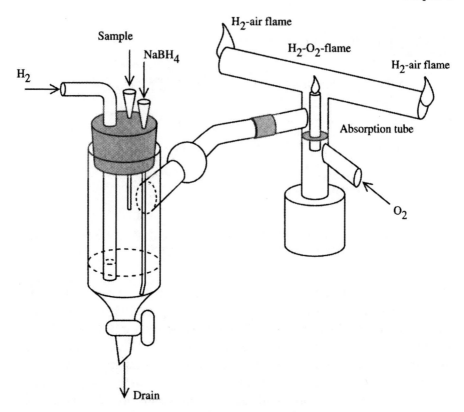

Figure 106 *A widely used batch type hydride generation system in which a small oxygen-hydrogen flame is burning inside an unheated quartz tube* (Adapted from: D. D. Siemer and L. Hagemann, Anal. Lett., 1975, **8**, 323)

3.5.2.1 Atomization of the Hydrides. The relatively easy atomization of the gaseous hydrides in heated quartz tubes and the increase in sensitivity with increasing atomizer temperature, led to a conclusion that atomization of the hydrides is a simple thermal dissociation. However, on the basis of a series later observations it has been shown that thermal dissociation is not the only atomization mechanism of the hydrides.

Dedina and Rubeska investigated the atomization of selenium hydrid in a cool oxygen-hydrogen flame burning in an unheated quartz tube. They concluded that atomization is brought about by the free radicals (OH and H) that are produced in the primary reaction zone of the flame. The concentration of H radicals is several orders of magnitude higher than the OH radicals. Thus, most probably atomization takes place via a two-step mechanism with the predominating H radicals:

$$SeH_2(g) + H(g) \rightarrow SeH(g) + H_2(g) \tag{73}$$

$$SeH(g) + H(g) \rightarrow Se(g) + H_2(g) \tag{74}$$

Correspondingly, reactions with OH radicals are also possible, but because of their low concentration these reactions are negligible. However, recombination must also be taken into account:

$$Se + H \rightarrow SeH \tag{75}$$

After a sufficiently large number of collisions with H radicals there is equilibrium between reactions (73), (74), and (75). The probability of the formation of free Se atoms from SeH_2 is proportional to the number of collisions with free H radicals and the efficiency of atomization should increase with increasing number of radicals. Analogous conclusions can be drawn for atomization of selenium hydride and other volatile hydrides in hydrogen diffusion flames.

In heated quartz tubes oxygen plays an active role in the atomization of gaseous hydrides. If measurements are performed with the exclusion of air (oxygen), all other hydrogen forming elements except bismuth exhibit either no measurable signal or distinctly reduced sensitivity at 1000–1100 K. With increasing temperature, sensitivity of all elements increases. In contrast, in the presence of oxygen maximum sensitivity is obtained at 1000 K and further increases in temperature do not bring further signal increases. When there is adequate oxygen and hydrogen (from the reaction of $NaBH_4$ with the acidified sample solution) in the quartz tube, it is conceivable that similar reactions to those in a fuel rich hydrogen flame take place.

Arsenic hydride (arsine) is not measurable when atomized in a pure argon atmosphere, even in a quartz tube heated up to 1300 K. The hydride is thermally decomposed, but its dissociation does not lead to free arsenic atoms. Probably the stable tetramer As_4 or dimer As_2 is formed. Atomization takes place as soon as hydrogen is mixed with the argon. Maximum sensitivity is attained at about 900 K when some oxygen is also introduced (argon with 1% oxygen).

Hydrogen is essential for the atomization of gaseous hydrides in a heated quartz tube and oxygen plays a supporting role at least at lower temperatures.

3.5.2.2 Gas-phase Interferences. Interferences in hydride generation method is discussed in Section 2.3.2. Virtually, all hydride forming elements interfere mutually. In some cases even very small concentrations of the interfering element cause severe signal depressions. The extent of the interference depends only on the concentration of the interfering element, not on the interferent-analyte ratio. After sufficient dilution of the sample solution these interferences do not appear.

The hydride forming elements are normally present at very low concentrations in the sample solutions. When a large excess of sodium tetrahydroborate is employed, no shortage of reductant should occur. It can thus be assumed that hydrides of all hydride forming elements present in the solution are formed and transported to the atomizer. Atomization of hydrides takes place via collisions with H radicals as explained above.

As^{III} interferes in the determination of selenium at a tenfold lower concentration (>0.01 mgl^{-1}) than As^{v}, which only shows an influence above 0.1 mg l^{-1}. Selenium in turn interferes much more in the determination of arsenic at concentrations above 0.001 mg l^{-1}. In addition, the interference is independent of the oxidation state of arsenic. In both cases, the degree of interference is only dependent on the concentration of the interferent. Selenium(IV) reacts the quickest, followed by arsenic(III), and arsenic(V) is the slowest. Selenium hydride volatilizes earlier than arsenic hydride. The concentration of H radicals needed for atomization in a heated quartz tube purged by an inert gas is not large and it is possible that a shortage of radicals can occur, especially when oxygen is largely excluded. Selenium hydride reaches the atomizer earlier than arsenic hydride. Radicals are consumed for atomization of selenium hydride and when arsenic hydride reaches the atomizer not enough radicals are available to cause the same degree of atomization that is achieved in the absence of selenium. This effect is not dependent on the appearance time of arsine, which explains why the influence of selenium on As^{III} and As^{v} is the same. Arsenic interferes less in the determination of selenium than *vice versa* and the degree of interference of As^{v} is less than that of As^{III}, which also supports the theory above.

The degree of interference of other hydride forming elements on the determination of antimony increases in the following order: Pb < Bi < As <Te < Ge < Se < Sn (Table 16). Lead virtually does not interfere and the influence of bismuth is very small and appears only at high concentrations. All other hydride forming elements interfere when their concentration in the sample solution is 100 μg l^{-1} or more. The interference can be eliminated by a suitable addition of some masking reagent as represented in Table 16.

3.5.3 Cold Vapour Technique

Determination of mercury by cold vapour AAS technique can be accomplished in several ways. Nowadays continuous flow and flow

Table 16 *Influence of various hydride-forming elements on the determination of antimony, and the effect of masking agents*
(Source: L. H. J. Lajunen, T. Merkkiniemi, and H. Häyrynen, Talanta, 1984, **31**, 709)

Interferent	Sb-conc. ($\mu g\,l^{-1}$)	Masking agent	Relative intensities of the Sb signals (%) at various interferent concentrations ($\mu g\,l^{-1}$)						
			0	100	250	500	1000	2000	4000
Pb	20	—	100	102	101	102	102	100	101
	50	—	100	99	99	100	100	100	99
Bi	20	—	100	100	99	93	93	89	87
	50	—	100	100	99	97	94	90	99
	20	CuSO$_4$	100	100	100	100	98	96	94
	50	CuSO$_4$	100	100	100	100	98	97	96
As	20	—	100	99	91	82	77	66	49
	50	—	100	99	92	86	78	67	50
Te	20	—	100	98	90	79	67	61	56
	50	—	100	98	91	80	69	63	57
	20	KI-aa*	100	99	98	95	93	96	97
	50	KI-aa*	100	100	99	97	93	95	96
Ge	20	—	100	98	90	76	62	50	40
	50	—	100	98	91	76	62	51	42
Se	20	—	100	99	86	66	54	43	35
	50	—	100	98	87	68	56	46	36
	20	CuSO$_4$	100	100	100	100	99	96	92
	50	CuSO$_4$	100	100	100	99	101	102	103
	20	KI-aa*	100	100	101	101	100	100	101
	50	KI-aa*	100	100	100	101	101	100	101
Sn	20	—	100	91	80	61	11	9	9
	50	—	100	93	87	67	12	8	8
	20	KI-aa*	100	100	103	99	103	102	101
	50	KI-aa*	100	103	102	104	103	102	96

*Ascorbic acid.

injection systems are coming increasingly popular, but also batch-type systems are still in use. Batch-type methods may be divided into two categories, dynamic or static methods. In an open dynamic system the liberated mercury is transported through the absorption cuvette with the aid of carrier gas flow. Hence a signal varying with time is obtained.

Earlier the closed dynamic method was very popular in the determination of low levels of mercury. In the simple closed dynamic method (Figure 107), air is circulated by a small pump through the reaction vessel, spray trap, and absorption tube with the longest path length and minimum volume possible. A length of 15 cm and a diameter of 0.75 cm would be suitable for many AAS instruments. The cuvette is usually made of glass with ultraviolet transmitting windows (quartz). After the determination, mercury is collected, for instance, in an active carbon trap. In addition, the total volume of the system should be kept in

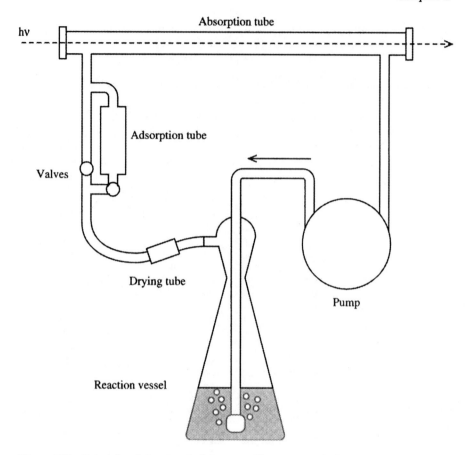

Figure 107 *Principle of the closed dynamic cold vapour method*

minimum; otherwise a reduced sensitivity is obtained when compared to that of open dynamic system.

The procedure is as follows: A clear sample solution, containing 0–300 mg of mercury, is transferred to the reaction vessel and the reductant is added. For example, 2 ml of 10% $SnCl_2$ solution. The bottle head is immediately inserted and the circulating pump switched on. Reduced mercury is liberated and swept through the absorption cuvette. The absorbance value rises to a plateau. When the constant absorbance value has been reached and recorded, mercury vapour is collected in a trap or allowed to escape through the extraction hood. When the absorbance returns to zero the next sample or standard can be inserted and reduced.

The repeatability of the dynamic system depends on the constancy of the total volume of the circulating air. Thus, the flask and sample solution volumes should be exactly the same during a particular analysis run.

The time required to liberate mercury increases with increasing sample volume and the increasing mercury content of the sample. In practice, a suitable sample volume is about 30–50 ml. For example, 50 µg of mercury may be liberated in 15 seconds from a sample volume of 30 ml, and 2 µg of mercury may be liberated in 10 seconds from a sample volume of 10 ml, whereas from 500 ml the time required is 4 minutes. When the circulating rates are too high the sample solution starts to foam, and when they are too small the liberation of mercury takes too long time.

The inorganic and organic mercury fractions of a sample may be determined in the following way: Inorganic mercury, Hg^{II} compounds, are reduced by 10% $SnCl_2$ solution. The liberated mercury vapour is circulated until the stable maximum absorbance value is reached. The vapour is then circulated through an active carbon trap until the absorbance returns to zero. After that a desired amount of a standard mercury solution is added to the reaction vessel and its absorbance reading recorded in the same way. In order to reduce organomercury compounds, $SnCl_2$–$CdCl_2$ solution is added to the reaction vessel. The calibration for the organomercury fraction is carried out by a standard methyl mercury solution.

In the static method, the reduction of mercury is performed as in the dynamic method. The mercury vapour is passed first through a vacuum flask, and then via a gas pipette to the absorption cuvette (Figure 108). In practice, the precision of the static method is better than that for the dynamic one since the mercury vapour is not circulated through the absorption cuvette. Hence the circulating rate does not affect the results, and it is not necessary to measure standards so often.

3.5.4 Semi-flame Methods

In semi-flame methods the atomization of the analyte takes place in a flame, but the pretreatment of the sample is outside the flame.

3.5.4.1 Delves Cup. The Delves cup method was first introduced by Delves for the determination of lead in blood samples. The method may be used for the determination of volatile elements such as Ag, As, Bi, Cd, Se, Te, Tl, and Zn.

The principle of the Delves cup method is shown in Figure 109. A sample is added to the nickel crucible. The thermal pretreatment of the sample (drying, ashing) may be performed on a hot plate or by radiation from the flame. By using hot plates, large sample series can be

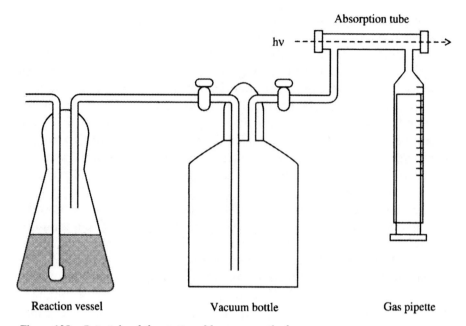

Figure 108   *Principle of the static cold vapour method*

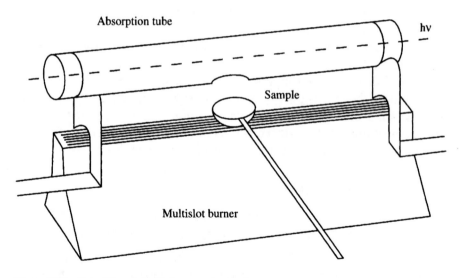

Figure 109   *Principle of the Delves cup method*

treated at the same time. A quartz absorption cell is supported in the optical axis with the multislot burner positioned about 2 cm below. A hole is made midway along the wall of the tube nearest the burner. The sample is introduced between the flame and the hole of the tube. The

Table 17 *Detection limits for some elements by the Delves cup method (A), and sampling boat method (B)*

Element	Absolute detection limit (ng)		Detection limit (mg l^{-1})	
	A	B	A^a	B^b
Ag	0.1	0.2	0.001	0.0002
As	20	20	0.2	0.02
Bi	2	3	0.02	0.03
Cd	0.005	0.1	0.00005	0.0001
Hg	10	20	0.1	0.02
Pb	0.1	1	0.001	0.001
Se	100	10	1	0.01
Te	30	10	0.3	0.01
Tl	1	1	0.01	0.01
Zn	0.005	0.03	0.00005	0.00003

Sample volume: ^{a}100 µl, ^{b}1 ml.

sample partially atomizes and the atoms are retained in the optical path by the tube. Non-specific absorption is caused during the atomization hence, background correction is always needed.

The sample amount required is very small and detection limits are lower than by conventional FAAS (Table 17). However, the reproducibility is not good and hence its most common use has been as a readily applied screening method.

3.5.4.2 Sampling Boat. The sampling boat method is even more simple the Delves cup method. A sample is added to a narrow, deep and long 'boat' (5 cm in length and 2 mm in width). The boat is made of tantalum because of its resistance to heat and good conductivity. A multislot burner preferred as it heats the boat more uniformly and produces a smooth flow of the combustion gases around the sample. A simple device is made to e the boat to be loaded about 10 cm away from the flame, then pushed close to the flame to dry the sample, and finally placed in the correct position within the flame.

The sampling boat method can be applied to the same volatile element as the Delves cup method (Table 17).

3.5.5 Indirect Methods

The method is called direct if the atomic absorption of the analyte is related to its concentration. Indirect methods in turn are based on the chemical reaction of the analyte with one or more species, where one should be a metal easily measurable by AAS. The reaction must be stoichiometric and a relationship between the atomic absorption of the metal and the concentration of the analyte must be established. Indirect methods are not common in routine analysis, because most of them are time-consuming and difficult to automate.

Indirect methods for the determination of anions and organic compounds may be classified into the three groups: (i) Methods based on the formation of compounds in solution; (ii) Methods based on the formation of compounds in the atomization cell; and (iii) Methods based on the formation of volatile compounds.

(i) Methods Based on the Formation of Compounds in Solution. These methods are based on precipitation, complex formation, redox, and heteropolyacid compound formation reactions.

An inorganic or organic anion is precipitated with a solution of a cation of appropriate concentration. The compound formed should have a very low solubility. An excess of the cation is used in order to obtain complete precipitation. The mixture is filtered, centrifuged, or decanted, and the cation is determined by AAS either in the filtrate or in the precipitate. These methods are specific only if the sample solution does not contain other species that might be precipitated.

The determination of the unreacted metal ion used as the precipitation reagent, in the supernatant solution, is more often employed than the determination of the precipitated metal after dissolution of the precipitate, as less manipulation is required. Calibration solutions are generally prepared from the anion standard solutions treated in the same way as the sample solutions.

Indirect methods for the determination of sulfur and chlorine by precipitation with metal cations and measurement by FAAS of the excess metal in solution are given in Table 18.

Several indirect methods for the determination of organic compounds by means of precipitation reactions have been proposed (Table 19). The organic compounds are precipitated with a metal salt solution, and the unreacted metal in the supernatant solution is determined by AAS.

In those methods, where the metal is determined in the dissolved precipitate, the amount of cation added is not critical, but more sample preparation steps are required.

Sulfur may be determined by oxidizing it first to the sulfate, which is then precipitated with barium chloride solution. The insoluble $BaSO_4$ is dissolved with basic EDTA solution, and the atomic absorption of barium is measured at 553.8 nm by FAAS.

The determination of chloride involves the precipitation of chloride with silver nitrate, dissolution of the precipitate in ammonia, and

Table 18 *Indirect determination methods for sulfur and chlorine by precipitation as insoluble metal sulfates or chlorides and measurement by FAAS of the excess of metal in solution*
(Source: M. Garcia-Vargas, M. Milla, and J. A. Perez-Bustamante, Analyst (London), 1983, **108**, 1417 and references therein)

Analyte/ sample	Precipitation medium	Precipitate	Separation method	Wavelength (nm)	Comments
S/Potable water	pH 2.2	BaSO$_4$	Filter	553.6	
S/Natural water	0.02 M HCl	BaSO$_4$	Decant	553.6	BaSO$_4$ is allowed to stand 18 h; RSD: 1–22% (508–13 µg SO$_4^{2-}$ ml^{-1}).
S/Plants	HNO$_3$, HClO$_4$, HCl	BaSO$_4$	Filter	553.6	Triple slot burner; recovery: 95–102%
S/Biological materials	0.1 M HNO$_3$	BaSO$_4$	Centrifuge	553.6	S is oxidized to SO$_4^{2-}$ by conc. HNO$_3$ at 250°C.
S/Water	25% ethanol	PbSO$_4$	Centrifuge	283.3	SO$_2$ is oxidized to SO$_4^{2-}$ by 3% H$_2$O$_2$.
Cl/Plants		AgCl	Filter	328.1	
Cl/Serum		AgCl	Centrifuge	328.1	

determination of silver in the resulting solution by FAAS using air-acetylene or air-hydrogen flames. Standard silver nitrate solutions in ammonia are used for calibration since bromide and iodide interfere in the determination of chloride.

Iodine in organic compounds may be determined by reducing it to iodide followed by precipitation with silver nitrate. AgI is dissolved in an excess of iodide solution and silver is measured by FAAS in the resulting solution.

The determination of sulfide involves ZnS precipitation in ammoniacal medium and dissolution of the precipitate in HCl. The method is practically free from interferences, as the sulfide is distilled as hydrogen sulfide.

In the methods which are based on metal complex formation, the analyte reacts with either one or two reagents giving rise to a charged or neutral complex compound or an ion-association complex. The metal is

Table 19   *Indirect determination of organic compounds via precipitation and measurement by FAAS of the metal excess*
(Source: M. Garcia-Vargas, M. Milla, and J. A. Perez-Bustamante, Analyst (London), 1983, **108**, 1417 and references therein)

Compound to be determined	Sample matrix	Precipitated compound	Measured metal	Comments
Oxalic acid	Urine	CaC_2O_4	Ca	Method can be used as a routine method
Phenylacetylene	Water	$AgC{\equiv}CPh$	Ag	
CHI_3, theobromine, sodium salicylate, mercaptobenzo-thiazole, potassium ethylxanthate, organic acids		Insoluble Ag-compounds	Ag	Method can also be employed to determine alkylated barbiturates and mercaptans
N, N'-Diphenyl guanidine (DFG)	$(DFG)_2CdI_4$	Cd	$Zn(SCN)_4^{2-}$ or $CdBr_4^{2-}$ can also be used	
Sulfonamides	Tablets, suspensions, injections	Insoluble Ag-compounds	Ag	A simple, selective, and precise method

then determined by AAS after filtration or solvent extraction. The metal complex formed must be very stable. Standards for calibration are prepared in the same way as the samples. Interference is caused by other anions present in the sample solution that might also be complexed with the metal cation.

The methods based on complex formation may be classified into two groups on the basis of the nature of the complex species formed: (i) Metal complexes with the analyte anion as ligand; (ii) Ion-pair compounds with the analyte anion as counter ion.

Cyanide is determined by measuring the atomic absorption of either silver, copper, or nickel in their corresponding cyano complexes, $Ag(CN)_2^-$, $Cu(CN)_3^-$, and $Ni(CN)_4^{2-}$. The most sensitive of these is based on the silver complex.

Sulfite can be determined indirectly with mercury. $Hg(SO_3)_2^{2-}$ is formed in the reaction of mercury(II) oxide (about 120 mg) and sulfite (about 120–850 µg) at pH 11. The atomic absorption of mercury is measured at 253.7 nm employing the air-acetylene flame. Iodide, thiosulfite, and thiocyanate interfere, but sulfate, nitrite, and cyanide do not interfere. A very sensitive method for the determination of sulfite or sulfur dioxide is based on the following disproportionation reaction:

$$Hg_2^{2+} + 2SO_3^{2-} \rightarrow Hg + Hg(SO_3)_2^{2-} \tag{76}$$

Metallic mercury formed during the reaction is determined by the cold vapour method. The detection limit is about 5 ng SO_3^{2-}.

Fluoride can be determined by means of an iron(III) thiocyanato complex extracted into isobutylketone. Iron is extracted back into the aqueous phase with the fluoride sample solution. The atomic absorption signal of iron is directly proportional to the concentration of fluoride (0.5–6 µg F^- ml^{-1}). EDTA can be determined by a similar technique. Copper is first extracted as the hydroxyquinolinato complex into isobutylketone, and then extracted back into the aqueous solution with the EDTA sample solution. In these methods the analyte anion must form a more stable complex compound with the metal ion than the ligand used for the first solvent extraction. These kind of methods may be applied for determination of various organic compounds such as polycarboxylic acids, amino acids, and alcohols which form stable metal complexes.

Cyanide, thiosulfite, sulfide, halides, and a number of organic anions can be determined by extracting their uncharged metal complexes into the organic phase, in which the atomic absorption of the central ion is measured. Methods based on the extraction are usually more sensitive than those presented above.

A very sensitive and selective method for the determination of cyanide is based on the extraction of cyanide as dicyanobis(1,10-phenanthroline)-iron(II) complex into chloroform. The chloroform is evaporated and the residue is dissolved in ethanol in which the iron signal is measured. The detection limit is 0.06 µg CN^- ml^{-1}.

A selective method for determining thiocyanide is based on the quantitative formation of the dithiocyanatodipyridine-copper(II) complex and its extraction into chloroform. The organic layer is sprayed into an air-acetylene flame and the copper signal at 324.7 nm is measured (detection limit 0.2 mg SCN^- l^{-1}). However, the aspiration of chloroform solution into the flame is not recommended since toxic combustion products may be formed. Hence, the organic phase should be evaporated almost to dryness, diluted with ethyl acetate, and the copper atomic signal measured as above (detection limit 0.05 mg SCN^- l^{-1}).

Chlorides have been determined at very low levels by extracting phenylmercury(II) chloride into chloroform, evaporating the organic phase to dryness, and dissolving the residue in ethyl acetate. The atomic absorption of Hg is then measured using an air-acetylene flame. The detection limit is very low, being about 0.015 mg Cl^- l^{-1}. Alternatively, isopropyl acetate may be used instead of chloroform, and the organic layer left after the extraction is sprayed into the flame. These two methods are more sensitive and less affected by the presence of other ions than other indirect AAS methods for determining chloride. However, iodide and bromide cause interference since they also react quantitatively with the reagent to form the corresponding phenylmercury(II) halides, which are then extracted into the organic phase. On the other hand, this may be used for determination of the total halide content of the sample, followed by stepwise elimination of iodide and bromide by chemical reactions in order to obtain the individual halide contents.

Table 20 lists some indirect methods for organic compounds which are based on the formation of uncharged metal complexes. The atomic absorption signal of the metal is measured either in the supernatant solution or in the organic phase after extraction.

The determination of perchlorate, nitrate, and iodide anions is based on the formation of an ion-association pair compound. An uncharged ligand forms a stable complex with the metal ion, which further forms an ion-association pair with the inorganic anion. The ion-association pair species is then extracted into the organic phase. 1,10-phenanthroline (phen) and 2,9-dimethyl-1,10-phenanthroline (neocuproine) are frequently used as organic ligands in these methods.

The determination of perchlorate or nitrate is based on the extraction of $[(neocuproine)_2Cu^I]X$ ($X = ClO_4^-$ or NO_3^-) into an organic phase and measurement of copper atomic signal at 324.7 nm in an air-acetylene flame. The perchlorate ion-association pair is extracted into ethyl acetate and the nitrate ion-association pair into MIBK. The method is suitable for solutions containing 0.5 to 5 mg ClO_4^- l^{-1}. The same method can also be employed for the determination of nitrites and nitrocompounds. Nitrite is first oxidized with cerium(IV) sulfate to nitrate and the nitrogroups to nitrate with potassium permanganate.

Iodide can be extracted as $[(phen)_3Cd^{II}I_2]$ ion-association pair into nitrobenzene and the atomic absorption of cadmium is measured at 228.8 nm in an air-acetylene flame. Phosphate, sulfate, cyanide, chloride, bromate, fluoride, and borate anions do not interfere, but periodate, perchlorate, chlorate, nitrate, nitrite, and bromide do. The calibration graph is linear between 10 and 40 mg l^{-1}.

A number of methods based on the formation of ion-association pairs have been proposed for the determination of organic compounds (Table 21).

Table 20 *Indirect determination methods for organic compounds by formation of neutral metal complexes and measurement by FAAS of the coordinated metal*
(Source: M. Garcia-Vargas, M. Milla, and J. A. Perez-Bustamante, Analyst (London), 1983, **108**, 1417 and references therein)

Compound to be determined	Measured metal	Description of the method
Esters and acid anhydrides	Fe	Hydroxamic acid formed is complexed with Fe^{III} and filtered off. Fe absorption signal is measured in the filtrate.
Ketones and aldehydes of low molecular mass	Cu	Determination is based on the Cu-thiosemicarbazone complex formation and its extraction into benzene. Ethanol is used for dilution.
Primary amines	Cu	Schiff's bases are formed, complexed with Cu, filtered off, and excess of Cu determined in the filtrate, or in the precipitate dissolved in HNO_3.
Secondary amines	Cu	Cu-dithiocarbamate complex is formed and extracted into MIBK.
CS_2	Cu	CS_2 is trapped in KOH and H_2S removed with Cd. Cu-xanthate is formed, filtered, and dissolved.
Anthranilic acid	Co	Co^{II}-anthranilic acid-bathophenanthroline complex is formed and extracted into MIBK.
Sodium diethyldithiocarbamate	Cu	Cu-diethyldithiocarbamate complex is formed and extracted into CCl_4
Free fatty acids	Co	Determination is based on the Co complex formation.

Table 21 *Indirect determination methods for organic compounds by formation of ion-association complexes and measurement by FAAS of complexed metal*
(Source: M. Garcia-Vargas, M. Milla, and J. A. Perez-Bustamante, Analyst (London), 1983, **108**, 1417 and references therein)

Compound to be determined	Measured metal	Ion-association compound	Organic solvent
Pentachlorophenol (PPh)	Fe	$[Fe(phen)_3]$ $(PPh)_2$	Nitrobenzene
Salicylic acid (H_2sal)	Fe	$[Fe(phen)_3]$ $(Hsal)_2$	Nitrobenzene
2-Hydroxynaphthoic acid (H_2naph)	Ni	$[Ni(phen)_2]$ $(Hnaph)_2$	Nitrobenzene
Benzylpenicillium (BP)	Cd	$[Cd(phen)_2]$ $(BP)_2$	
Anionic detergents (AD)	Cu	$[Cu(phen)_2]$ (AD)	MIBK
Aliphatic amines (AA)	Co	$[Co(SCN)_4]$ $(AA)_2$	Benzene

Heteropolyacid compounds. Some anions react with molybdate and other ions in solution, yielding heteropolyacids of definite chemical composition. These heteropolyacid compounds are then usually extracted into an organic phase. The metal of the compound is determined by AAS either in the organic phase or in aqueous solution after back-extraction.

Heteropolyacids contain 10, 11, or 12 molybdenum atoms per phosphorus atom. Indirect methods for the determination of phosphate are given in Figure 110. The most suitable organic solvents for extracting heteropolyacids are oxygen-containing solvents such as alcohols (n-butanol, isobutanol, hexanol, heptanol), ketones (MIBK), and esters (isobutyl acetate, butyl acetate).

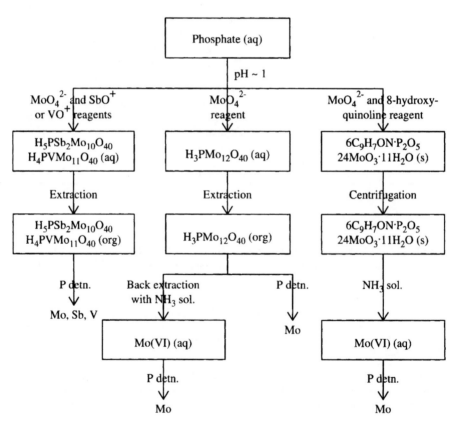

Figure 110 *Indirect determination of orthophosphates based on the formation of heteropolyacids*
(Adapted from: M. Garcia-Vargas, M. Milla, and J. A. Perez-Bustamante, Analyst (London), 1983, **108**, 1417)

In the back-extraction, molybdophosphoric acid is first extracted into either ester, ether, or their mixture. Then molybdenum is extracted back into the aqueous phase with ammonia solution. The atomic absorption of molybdenum is measured in the aqueous phase by FAAS or GF-AAS.

The organic phase may also be evaporated to dryness and the residue redissolved in nitric acid. If back-extraction is employed, standards can be prepared directly in aqueous solutions.

Redox reactions. Oxidizing or reducing substances may be determined indirectly by means of the metal ions that can be oxidized or reduced by the substances involved. The reduced or the oxidized form of the metal ion is then separated for the atomic absorption measurement. Generally, these methods are not very selective, because the sample must not contain any other oxidizing or reducing agent than the one to be determined.

Inorganic anions such as iodide, iodate, nitrate, and chlorate have been determined on the basis of redox reactions. Iodides have been determined by their reaction with chromium(VI) in an acidic medium. The unreacted Cr^{VI} is then extracted into MIBK from a 3 M HCl solution. The iodide concentration is quantitatively related to the increase of the Cr atomic signal in aqueous solution or the decrease of the Cr signal in the organic phase. Selenium(IV) is reduced by iodide to elemental selenium. The atomic absorption signal of selenium is then measured in the selenium precipitate.

Iron(II) is oxidized to iron(III) by iodate ions. Iron(III) may be then extracted into diethyl ether from 9 M HCl solution. Iron(III) is determined in organic phase by FAAS.

Nitrates have been determined by reduction with cadmium metal in a dilute HCl solution. The dissolved cadmium is then determined by FAAS (228 nm, an air-acetylene flame):

$$2NO_3^- + 4Cd + 10H^+ \rightarrow N_2O + 4Cd^{2+} + 5H_2O \qquad (77)$$

Chlorate is reduced to chloride by iron(II). Chloride ions formed are then precipitated with silver nitrate. The precipitate is dissolved in ammonia and the Ag atomic signal is measured.

Organic compounds such as aldehydes, alcohols, aliphatic, and aromatic 1,2-diols, nitrocompounds, and sugars have been determined by using two types of redox reactions. The analyte may be reduced with a metal ion in solution and the metal precipitate formed dissolved in nitric acid, and determined by AAS. Alternatively, the unreduced metal may be determined. The other type of method is based on the oxidation of the analyte with periodic acid or potassium permanganate, and addition of metal ion solution. The added metal ion forms, with the oxidized form of the analyte, either an insoluble compound or an uncharged metal complex which can be extracted into the organic phase. The atomic absorption of the metal ion is determined after dissolution of the precipitate or solvent extraction.

Polyhydric alcohols react with HIO_4 to yield iodate which is precipitated with silver nitrate, and the precipitation is dissolved in ammonia. Detection limits for glycerol, lactose, and glucose are 0.16, 0.82, and 0.28 mg, respectively.

Sugars have been determined with copper. Cu^{II} reacts with sugars in alkaline medium to yield Cu_2O. Copper(I) oxide is removed and unreacted copper(II) is determined.

(ii) Methods Based on the Formation of Compounds in the Atomization Cell. Some anions form refractory compounds with metals at the atomization cell temperature. These methods are divided into two groups depending on whether the absorption measurements are monitored on either atoms or molecules. These determinations are based on the reaction between the analyte anions and the metal cations taking place in the atomization cell, not in the sample solution. Therefore, these methods are often more rapid as no prior separation of the analyte species is required.

Molecular absorption spectrometry (MAS) is based on the formation of stable, volatile two-atomic compounds in the atomization cell, and the molecular absorption of the compound formed is then measured at the absorption band by means of a continuum or a line-like radiation source. These methods are described in the next Section (3.5.6).

The formation of refractory compounds in the atomizer may suppress or enhance the atomic absorption signal of a given metal. These methods are rapid, but other species present in the sample that might suppress or enhance the absorbance of the metal added cause interference. In order to overcome these interferences, the sample should be treated previously with a cation- or anion-exchanger. The determinations may be performed by adding a constant concentration of the metal ion to the sample solution or by an atomic absorption titration.

Constant concentration of the metal added. The calibration graph is constructed by using a constant metal concentration in the presence of increasing amounts of the analyte anion. The slope of the calibration graph is negative when the absorbance of the metal is presented as a function of the added analyte concentration. The analyte concentration is the sample solution determined by adding the same metal concentration as in the calibration graph, measuring the atomic absorption of the metal, and relating this absorbance to the calibration graph.

Fluorine content of fluorine-containing compounds may be determined indirectly by means of sodium. Fluorine and sodium react in the atomizer at about 1100 K to yield thermally stable sodium fluoride which suppresses the atomic absorption signal of sodium. Chloride, bromide, and iodide ions do not interfere, and the method is very sensitive (the detection limit being about 0.8 ng).

Phosphate, sulfate, and fluoride ions as well as many organic compounds may be determined by means of alkaline earths. For example, calcium has been used for the determination of phosphate and sulfate. Phosphate may also be determined by strontium. Magnesium has been employed for the determination of fluoride and calcium for the determination of glucose.

A number of varying methods based on the formation of stable transition metal compounds have been proposed. For example, fluoride and ammonia may be determined by means of zirconium.

Atomic absorption titration. These methods are divided into two groups: (i) atomic absorption inhibition titrations; and (ii) atomic absorption inhibition-release titrations.

Atomic absorption titration involves the addition of the metal titrant solution to a stirred solution of the anion from which metal cations have been removed by ion exchange. The titrated solution is simultaneously aspirated into the flame, and the atomic signal of the metal is monitored.

Typical calibration graphs are presented in Figure 111. Sulfate has been determined by titrating with magnesium ions (Figure 111A). The sulfate ion concentration may vary between 1 and 20 μg l^{-1}. Phosphate and silicate interfere, and these ions should be removed before the determination.

Magnesium can be used for the determination of ortho-, pyro-, tri-, tetra-, and hexaphosphates. The shape of the titration curve for orthophosphate (Figure 111B) suggests the formation of refractory compounds with 2, 3, and 4 magnesium atoms for each phosphorus atom. At any of these points the magnesium-phosphate ratio remains constant, and all these points may serve as an end-point for the titration, even in the presence of sulfate.

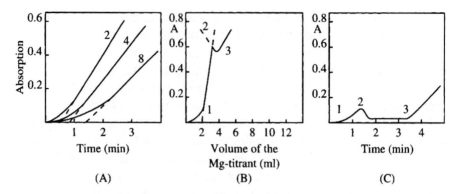

Figure 111 *A: Determination of sulfate (2, 4, and 8 µg ml⁻¹) by titration with magnesium solution (50 mg l⁻¹, flow rate 1.49 ml min⁻¹). (Adapted from: R. W. Looyenga and C. O. Huber, Anal. Chim. Acta, 1971, 55, 170). B: Determination of phosphate (8 µg ml⁻¹) by titration with magnesium solution (200 µg ml⁻¹). (Adapted from: W. E. Crawford, C. I. Lin, and C. O. Huber, Anal. Chim. Acta, 1973, 64, 387). C: Determination of silicate (1.0 µg ml⁻¹), phosphate (4.0 µg ml⁻¹), and sulfate (2.0 µg ml⁻¹) by titration with magnesium solution (50 µg ml⁻¹, flow rate 2.03 ml min⁻¹)*
(Adapted from: C. I. Lin and C. O. Huber, Anal. Chim. Acta, 1972, 44, 2200)

The simultaneous titration of silicate, phosphate, and sulfate with magnesium is a very rapid, sensitive, and selective method for these anions (Figure 111C). The titration curve shows three distinct rectilinear segments (points 1, 2, and 3) with three different slopes, corresponding to silicate, silicate and phosphate, and silicate, phosphate, and sulfate, respectively.

Magnesium is selected as the titrant metal due to its atomic absorption sensitivity and large inhibition that can be brought about by anions forming magnesium refractory compounds.

(iii) Methods Based on the Formation of Volatile Compounds. A specific method for the determination of fluoride is based on the formation of gaseous SiF_4. The atomic absorption signal of silicon is measured at 251.6 nm using either a dinitrogen oxide-acetylene flame (detection limit 30 µg of F^-) or a graphite furnace (detection limit 0.17 µg of F^-).

A rapid, but not very selective and sensitive method for the determination of chloride (50–200 mg Cl^- l⁻¹) is based on the formation of gaseous CuCl.

By heating solutions containing iodide and mercury(II) nitrate in a graphite furnace, volatile mercury(II) iodide, HgI_2, is formed. Two absorption peaks will appear because the decomposition temperature of mercury(II) iodide is higher than that of mercury(II) nitrate. The interference caused by cyanide, thiosulfate, and sulfide ions may be avoided by adding H_2O_2 to the solution.

Indirect methods for metallic elements. Some metals are difficult to determine by direct atomic absorption methods owing to their low sensitivity or poor atomization. Various indirect methods have been proposed for these metals. Most of the methods are based on the formation and separation of heteromolybdophosphates (Table 22).

3.5.6 Molecular Absorption Spectrometry with Electrothermal Vaporization (ETV-MAS)

The direct determination of non-metals, such as sulfur, phosphorus, and halogens by conventional AAS methods is either impossible or very insensitive due to disadvantageous characteristics of these elements (Table 23). The most useful spectral lines of the non-metals are located in the vacuum UV region. In addition, their compounds are often very volatile and difficult to dissociate, which makes them difficult to atomize and excite.

The tendency of the non-metals to form stable, volatile, two-atomic molecules was first applied in emission spectrometry (Table 24). In most cases, cool diffusion flames were employed as spectral sources.

Table 22 *Indirect determination methods for metallic elements*

Analyte	Measured metal	Description of the method
Al	Fe	Al enhances the absorption signal of Fe. The change in the signal is directly proportional to the Al concentration.
Ge	Mo	Ge-molybdophosphoric acid is extracted first into a mixture of 1-pentanol and diethylether and then extracted back into ammoniacal aqueous solution. The Mo absorption is measured in the aqueous phase.
Nb	Mo	Niobium-molybdophosphoric acid is extracted into isobutyl acetate or butanol.
Tl	Mo	$Tl_2HPOMo_{12}O_{40}$ is precipitated, filtered, and dissolved in 0.1 M NaOH solution.
Th	Mo	Thorium molybdophosphoric acid is extracted into isobutyl acetate.
Ti	Mo	Titanium molybdophosphoric acid is extracted into CCl_4 or n-butanol.
V	Mo	Vanadium molybdophosphoric acid is first extracted into n-pentanol or diethylether and then extracted back into aqueous solution buffered with an NH_3-NH_4Cl buffer.

Table 23   *Resonance line wavelengths and ionization and dissociation energies*
for non-metals
(Source: K. Dittrich and B. Vorberg, Chemia Analityczna
(Warsaw), 1983, **28**, 539)

Element	Wavelength of the resonance line (nm)	Ionization energy (eV)	Dissociation energy (eV)
B	249.7	8.3	2.8
C	165.6	11.3	6.2
Si	253	8.1	
N	120.0	14.5	9.8
P	179	10.5	4.9
As	193	9.8	3.4
O	130.0	13.6	5.1
S	181	10.4	4.4
Se	196	9.8	3.4
Te	214	9.9	2.7
F	95.5	17.4	1.4
Cl	135	13.0	2.4
Br	149	11.8	2.0
I	178	10.4	

Table 24   *Determination methods for non-metals based on the formation and*
emission of diatomic molecules

Method	Emitting molecule	Analyte element
Flame-AES	InF, GaF	F
	InCl, GaCl	Cl
	InBr, GaBr	Br
	InI	I
	S_2	S
	PO	P
Arc-AES	GaF	F
	CN	N

In 1977 Dittrich and in 1978 Fuwa proposed, independently, that
non-metals could be determined by molecular spectrometry at high
temperatures with electrothermal vaporization (ETV-MAS). The
method is relatively simple and allows determinations at sub p.p.m.-
levels. Sample and reagent solutions are introduced into the graphite
tube as a mixture or one after another. The reproducibility is usually
better when the sample and reagent solutions are mixed together before
introduction. However, if precipitation occurs by mixing the solutions,
then they must be introduced separately into the atomizer. The sample
is dried and ashed like in GF-AAS. After the thermal pretreatment

steps, the conditions of the furnace are chosen so that the two-atomic analyte compound will be evaporated but not dissociated. This step corresponds to the atomization stage in GF-AAS.

Molecular absorption methods described in the literature are based on the use of continuum light sources (deuterium lamp with a thermal cathode or hollow cathode) or line-like radiation sources (hollow cathode lamps). Measurements using a continuum light source are carried out with a dual-channel instrument. The other channel is needed for the background correction. With a line-like radiation source, a conventional AA spectrometer can be used. In this technique the non-specific absorption is measured with a continuum radiation source.

In the sample preparation, introduction, and drying, several points must be considered. The furnace heating program is dependent on the way the sample and reagent solutions are introduced into the atomizer. During the drying stage analyte molecules can be evaporated to some extent. For example, if the solution is made alkaline with an alkali metal hydroxide, an excess of the alkali metal may decrease the absorbance signal according to the following gas phase reaction:

$$M_1X + M_{alk} \rightarrow M_1 + M_{alk}X \tag{78}$$

Also a large excess of reagent solution may reduce the absorption signal due to the formation of a non-volatile compound with the analyte anion. If the reagent (metal ion) forms an insoluble compound with the analyte anion, these solutions must be introduced one after another into the graphite tube. Good results have been obtained if the sparingly soluble compound can also be evaporated (for example, TlI and AgCl).

The thermal pretreatment stage must be carefully controlled in order to avoid losses of the analyte. It must be ensured that neither the cation nor anion of the volatile compound evaporate during this stage. However, the evaporation of the anion is hard to prevent, because the anion is generally much more volatile than the cation.

A special problem arises with metal salt hydrates which may hydrolyse at a high temperature and, thus interfere with the measurements according to the following reaction:

$$MX_n \cdot mH_2O \rightarrow MX_{n-1}OH \cdot (m-1)H_2O + HX \tag{79}$$

Salts of weakly basic metals, such as Al or Ga, containing crystal water molecules decompose according to this reaction. An oxide will be formed and the anion will volatilize. The evaporation of the anion can be diminished with addition of some strong bases.

In order for the diatomic molecule (MX) to form and its absorbance to be measured, the metal, M and the anion, X must evaporate at the same time. The molecule is thermally stable enough when its dissociation energy (E_d) is 4 to 6 eV. However, the detection of less stable molecules ($E_d < 4$ eV), is also possible under suitable conditions. To ensure the maximum formation of the desired molecule, a large excess of the metal reagent is used. In the most uncomplicated case the metal cation and non-metal anion form a solid compound during the thermal pretreatment, which then will volatilize during the evaporation stage. This has been observed for alkali and alkaline earth metal halides and partly for the silver and thallium halides. In other cases, the metal ion and non-metal anion volatilize separately. Because of the larger volatility of the anion, matrix modification is frequently employed to prevent the evaporation of the anion before the metal cation. Alkaline earth metals are used as matrix modifiers as they are basic, less volatile, and form stable compounds with the tube carbon.

In order to achieve good results, several parameters should be optimized in ETV-MAS. This technique involves more matrix effects than conventional AAS techniques.

In Table 25, the dissociation energies and wavelengths of the strongest absorption bands of the diatomic aluminium, gallium, indium, and thallium halides are given. Figure 112 shows the absorption spectra of AlCl and AlBr.

The observed absorbance is strongly dependent on the metal-reagent concentration in the atomizer. Compounds with small dissociation energies need a large excess of the metal reagent. Sodium also significantly affects the results, especially when the reagent used is aluminium. Aluminium is the least basic element of the boron group elements. Sodium diminishes the hydrolysis of the aluminium halides during the thermal pretreatment stage. In the case of indium, the effect of sodium is much less because indium is a strongly basic element. In addition, sodium decreases the InF signal because the dissociation energies of InF and NaF are similar. Hydrolysis increases the AlCl absorption signal. The increase in the signal is due to the diminished hydrolysis during the pretreatment stage.

Each halide interferes with the MAS determinations of another halide, because every halide forms stable compounds with the metal reagent used. For example, when fluorine is determined with aluminium, gallium, or indium, all other halides have an influence on

Table 25 *Dissociation energies and absorption wavelengths for diatomic halides (MX) of the Group 13 elements*
(Source: K. Dittrich and B. Vorberg, Chemia Analityczna (Warsaw), 1983, **28**, 539)

Metal	Dissociation energy (eV)				Wavelength (nm)			
	F	Cl	Br	I	F	Cl	Br	I
Al	6.8	5.1	4.1	—	227.5	261.4	279.0	—
Ga	6.0	4.9	4.2	3.5	211.4	248.2	268	—
In	5.3	4.5	4.2	3.4	233.9	267.2	284.5	318
Tl	4.4	3.8	3.4	2.9	224	251.9	266.8	202

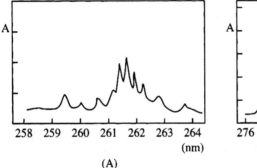

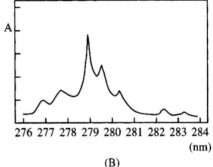

(A)   (B)

Figure 112 *Molecular absorption spectra of AlCl (A) and AlBr (B)*
(Adapted from: K. Tsunoda, K. Fujiwara, and K. Fuwa, Anal. Chem., 1978, **50**, 861)

Table 26 *Detection limits and interference caused by other halogens for the MAS determination of halogens*
(Source: K. Dittrich and B. Vorberg, Chemia Analityczna (Warsaw), 1983, **28**, 539)

Molecule	Absolute DL (ng)	Allowed excess of other halogens			
AlF	0.4	—	5×10^4	10^5	10^5
GaF	0.8	—	5×10^3	10^4	5×10^4
InF	1.1	—	10^4	10^4	5×10^4
MgF	2.4	—	2×10^4	4×10^4	4×10^4
AlCl	1.6	10^4	—	10^3	10^3
GaCl	9.0	10^2	—	5×10^2	5×10^3
InCl	3.0	10^5	—	10^3	10^3
MgCl	5.8	10^3	—	5×10^2	10^2
AlBr	18	10^2	5×10^2	—	10^3
InBr	130	10^2	10	—	10
TlBr	50	5×10^2	10	—	10^2
TlI	80	5×10^2	10^2	10	—

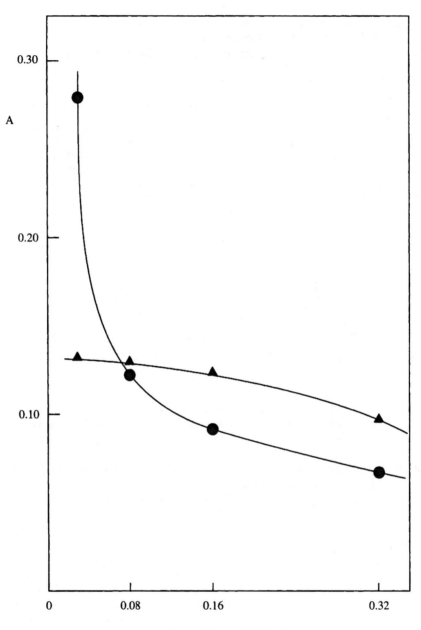

Spectral band width (nm)

the signal intensity. Table 26 summarizes the detection limits and interference effects of Al, Ga, In, and Tl halides.

When using a continuum radiation source, the observed absorption signal is dependent on the slit width (Figure 113). Some cations, such as Ni^{II}, Fe^{III} and Co^{III}, improve the sensitivity and decrease the background absorbance in the case of aluminium monohalides (Figure 114). The non-specific background absorption is mainly due to the formation of AlO.

Figure 115 shows the suitability of five hollow cathode lamps to the determination of chlorine by aluminium monochloride. The strongest signal is obtained with the Pb hollow cathode lamp at 261.4 nm. Of the

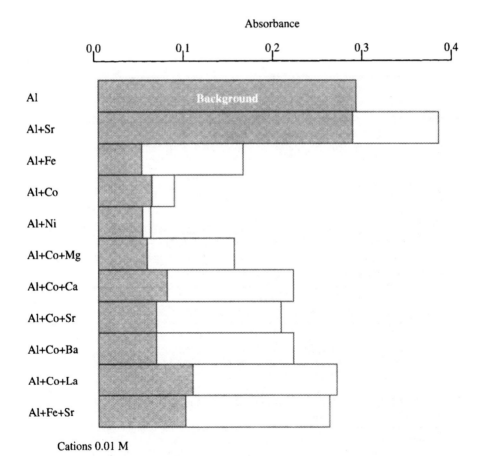

Cations 0.01 M

Figure 114 *Effect of cations on the AlCl absorption intensity in the graphite furnace* (Adapted from: K. Tsunoda, K. Fujiwara, and K. Fuwa, Anal. Chem., 1978, **50**, 861)

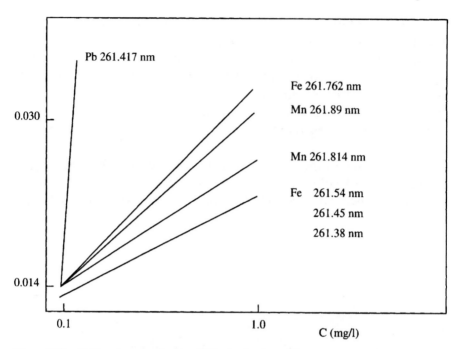

Figure 115 *Calibration graphs for AlCl obtained at different atomic absorption lines using hollow cathode lamps of metallic elements*
(Adapted from: P. Parvinen and L. H. J. Lajunen, Spectrosc. Lett., 1989, **22**, 533)

Table 27 *MAS methods for determination of non-metals (1%-abs. = characteristic concentration or mass, LDR = upper limit of linear dynamic range, RSD = relative standard deviation, HCL= hollow cathode lamp)*

Analyte element	Molecule	Radiation source	1%-abs.	LDR	RSD (%)	References
F	AlF	H₂-HCL	0.021 ng	0.6 ng	2	a
F	AlF	Pt-HCL	0.042 ng	2 ng	2	b
Cl	AlCl	H₂-HCL	0.02 mg l⁻¹	1 mg l⁻¹	2–4	c
Cl	AlCl	Pb-HCL	0.03 mg l⁻¹	3 mg l⁻¹	3	d
Br	AlBr	H₂-HCL	5.5 ng	100 ng		a
Br	AlBr	H₂-HCL	1.9 ng	100 ng	7	c
S	InS	H₂-HCL	3.1 ng			e

[a] K. Tsunoda, K. Fujiwara, and K. Fuwa, Anal. Chem., 1977, **49**, 2035.
[b] K. Tsunoda, K. Chiba, H. Haraguchi, and K. Fuwa, Anal. Chem., 1979, **51**, 2059.
[c] K. Tsunoda, K. Fujiwara, and K. Fuwa, Anal. Chem. 1978, **50**, 861.
[d] P. Parvinen and L.H.J. Lajunen, Spectrosc. Lett., 1989, **22**, 533.
[e] K. Fuwa, H. Haraguchi, and K. Tsunoda, 'Recent Advances in Analytical Spectroscopy', ed. K. Fuwa, Pergamon Press, Oxford, 1981, p. 199.

Table 28 *ETV-MAS applications*

Analyte/ molecule	Sample matrix	Certific. or calculated value	Obtained value	References
F/AlF	SRM 1571 (NBS: Orchard leaves)	4 µg g^{-1}	3.8 µg g^{-1}	a
F/AlF	Sodium acetate	0.62 ng	0.68 ng	a
F/AlF	o-Fluorobenzoic acid	0.59 ng	0.63 ng	a
F/AlF	Trifluoroacetic acid	0.52 ng	0.56 ng	a
F/AlF	SRM 2671 (NBS*: urine)	7.1 µg ml^{-1}	7.1 µg ml^{-1}	e
Cl/AlCl	SRM 1571 (NBS: Orchard leaves)	690 µg g^{-1}	630 µg g^{-1}	b
Cl/AlCl	SiO$_2$ (46%), Al$_2$O$_3$ (38%), Fe$_2$O$_3$ (1.1%), MgO (0.5%), K$_2$O+ Na$_2$O (0.36%)	0.503%	0.500%	c
Br/AlBr	p-Bromobenzoic acid	25.5 ng	28.2 ng	b
Br/AlBr	Sodium 2-bromosulfonate	25.6 ng	25.8 ng	b
Br/AlBr	KBrO$_3$	24.3 ng	26.4 ng	b
S/InS	KSCN	57.4 ng	58.8 ng	d
S/InS	Na$_2$SO$_3$	48.0 ng	45.8 ng	d
S/InS	L-cysteine	51.1 ng	47.6 ng	d

*The National Bureau of Standards (NBS) changed its name to the National Institute of Standards and Technology (NIST) in 1988.

[a] K. Tsunoda, K. Fujiwara, and K. Fuwa, Anal. Chem. 1977, **49**, 2035.
[b] K. Tsunoda, K. Fujiwara, and K. Fuwa, Anal. Chem. 1978, **50**, 861.
[c] P. Parvinen and L. H. J. Lajunen, Spectrosc. Lett., 1989, **22**, 533.
[d] K. Fuwa, H. Haraguchi, and K. Tsunoda, 'Recent Advances in Analytical Spectroscopy', ed. K. Fuwa, Pergamon Press, Oxford, 1981, p. 199.
[e] K. Tsunoda, H. Haraguchi, and K. Fuwa, 'Fluoride Research 1985, Studies in Environmental Science', ed. K. Tsunoda and M.-H. Yu, Vol. 27, Elsevier, Amsterdam, 1986, p. 15.

other halides, only fluoride interferes strongly with the determination when a line-like radiation source is employed.

Tables 27 and 28 summarize various ETV-MAS techniques and applications.

CHAPTER 4

Flame Atomic Emission Spectrometry

The majority of commercial atomic absorption spectrometers permit both flame atomic absorption and flame atomic emission measurements to be performed. Thus, flame AES is no longer considered as an independent instrumental technique, except for the determination of sodium and potassium (as well as calcium or lithium) in biological samples by flame photometers.

The excitation energy is inversely proportional to the wavelength. According to the Maxwell-Boltzmann law (Equation 18) the number of excited atoms in the flame decreases exponentially with increasing excitation energy and decreasing temperature. In the case of atomic absorption, the sensitivity of the method is directly proportional to the number of atoms in the ground state. In contrast, the sensitivity of atomic emission will increase with an increasing number of atoms in the excited state. In addition, the lifetime of the excited atoms is also an important factor in flame AES. The ratio of the excited atoms to the ground state atoms is more unfavourable for flame AES in the short wavelength range of the spectrum than in the long wavelength range (Figure 9). Flame AAS is more sensitive than flame AES when the excitation potential is greater than 3.5 eV.

Flame AAS and flame AES should be considered as complementary methods to each other. The flame AES technique is especially sensitive for alkali metals. Table 29 lists detection limits for flame AES, flame AAS, and ICP-AES for elements sensitive for flame AES measurements. In addition, by using hot, premixed flames (*i.e.* dinitrogen oxide-acetylene flame), emission measurements are as free of interferences as absorption measurements. On the other hand, small changes in the temperature of the flame may cause considerable changes in the number of excited atoms, whereas the number of atoms in the ground state remains almost unaltered. However, if the stoichiometry of the flame is

Table 29 *Instrumental detection limits ($\mu g \, l^{-1}$) attainable with flame AES, flame AAS, and ICP-AES*

Element	Flame AAS	Flame AES	ICP-AES
Al	10	10	20
Ba	1	10	0.5
Ca	0.1	1	0.1
Cr	5	2	5
Cs	8		
Cu	10	1	3
In	5	20	20
K	3	1	15
Li	0.03	3	3
Mg	5	0.1	0.2
Mn	5	2	1
Na	0.1	0.5	0.2
Rb	0.3	2	15
Sr	0.1	2	0.2
V	10	40	2

altered, the chemical properties are also changed which can strongly affect the production of free atoms. Thus, alterations of the flame stoichiometry affect both methods to about the same degree.

No additional light sources (hollow cathode lamps or EDLs) are required in flame AES, which can be regarded as an advantage of this technique. The most important difference between flame AAS and flame AES is spectral interferences due to overlapping of spectral lines in flame AES which are unusual in flame AAS. Essentially, this is a problem with the monochromator, and many such interferences may be avoided when monochromators having spectral bandwidths of 0.03 nm or better are used. In addition, background emission from the flame and sample matrix may cause interference in flame AES. Due to the strong background emission, detection limits in the presence of complex matrices may be considerably poorer in flame AES than in flame AAS. The detection limits given in Table 29 are instrumental detection limits which are valid only for pure aqueous solutions.

CHAPTER 5

Plasma Atomic Emission Spectrometry

In plasma atomic emission spectrometric methods the sample is introduced into a plasma source, where it is evaporated, and dissociated into free atoms and ions, and further additional energy is supplied to excite the free atoms and ions to higher energy states. Thus, the plasma cell is both the atomization and excitation source. A plasma is partially ionized gas, which remains macroscopically neutral, and is a good conductor of electricity. The high temperature of the plasma and the dissociation of the analyte compound to atoms and ions and their excitation are produced by collisions with other particles, mainly with free electrons. The excited state is unstable and the atom or atomic ion loses its excess energy either by collisions with other particles, or by a radiative transition to a lower energy level. The resulting radiation is called spontaneous emission of radiation. The AES methods are based on these spontaneous emission spectra. These spectra originating from a plasma source are very complex. The source contains many kinds of atoms and ions originating from the sample and gas streams, and the emission spectrum of each particle present in the cell has many lines. Therefore, a monochromator with a good resolution power and possibility for background correction is required for the spectrometer in AES.

The wavelengths of the emission lines are characteristic of the elements present in the plasma source. The detection of radiation at particular wavelengths can be applied to the qualitative analysis of the sample and the intensities measured at these wavelengths to the quantitative analysis of the analyte elements.

Most analytical plasma sources are electrical gas discharges at atmospheric pressure, usually in argon or in another inert gas. The various plasmas look like flames, but their temperatures are significantly higher-normally more than 5000 K in the viewing zone.

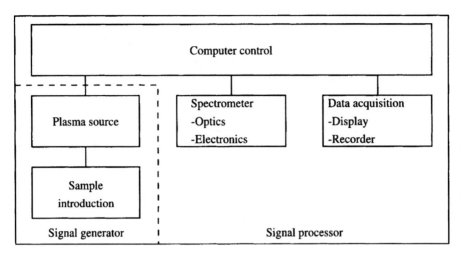

Figure 116 *Basic units of a plasma atomic emission instrument*

The analysis of liquids is rather difficult using classical spectroscopic sources (*i.e.* arcs and sparks), but various plasma sources permit the direct and easy analysis of liquids.

5.1 INSTRUMENTATION FOR PLASMA ATOMIC EMISSION SPECTROMETRY

The plasma atomic emission instruments consist of two main units (Figure 116): (i) the signal generator, and (ii) the signal processor. The signal generator consists of the plasma source and sample introduction system (an autosampler, pump, nebulizer, and a tube for sample introduction), and the signal processor comprises the optics and electronics and the data acquisition unit.

Various functions of the instrument and data acquisition are carried out by a micro computer. For example, in a scanning spectrometer the analyses are carried out automatically according to a preset program, which permits the rapid successive determination of each analyte element with optimum operating conditions.

5.2 PLASMA SOURCES

An ideal excitation source for AES should have the following features: (i) high specificity; (ii) high selectivity; (iii) high sensitivity; (iv) high accuracy; (v) high precision; (vi) capacity for multi-element determinations; (vii) ease of operation; (viii) free of matrix interferences.

Classical excitation sources in atomic emission spectroscopy do not meet these requirements. Flame, arc, and spark all suffer from poor stability, low reproducibility, and substantial matrix effects. However, the modern plasma excitation sources, especially ICPs, come very close to the specification of an ideal AES source.

The analytical plasmas are classified according to the method of power transmission to the working gas. There are three dominant types of plasma source in use today: (i) Inductively coupled plasmas, ICPs; (ii) Direct current plasmas, DCPs (current carrying DC plasmas and current-free DC plasmas): (iii) Microwave plasmas (microwave induced plasmas, MIPs, and capacitively coupled microwave plasmas, CMPs).

5.2.1 Inductively Coupled Plasmas

The most common plasma source is the inductively coupled plasma (ICP) torch. It consists of three concentric tubes, mostly made of quartz (Figure 117). The tubes are designed 'outer tube', 'intermediate tube', and ´inner or carrier gas tube'. Around the torch is a two or three turn induction coil, to which radio-frequency (RF) energy is applied. The coil is water-cooled. Flowing gases are introduced into the torch and the RF-field is switched on. An intense oscillating magnetic field around the coil is developed in which the lines of force follow elliptical closed paths (Figure 118). The magnetic field induces an electric field in the coil region.

When seed electrons and ions are introduced with a Tesla discharge, these start oscillating within the field and, provided that the magnetic field strength is high enough, a plasma will form. This results in great heating and an avalanche of ions. The ionized plasma acts as a shorted secondary winding of a transformer.

An inductively coupled plasma is maintained by inductive heating of the flowing plasma gas. The magnetic fields generated by the RF-currents in the coil produce high frequency, circular currents in the plasma (Figure 118). The temperature inside the coil is very high (about 10000 K), so a vortex gas flow between the outer and intermediate tubes must he used in order to prevent the torch from melting.

The sample is injected as an aerosol through the inner tube. The reactions when the sample mist passes through the plasma are shown in Figure 119. Even refractory compounds are easily decomposed in the ICP. Many physical characteristics of the ICP make it an ideal spectroscopic source. Sample particles experience a temperature of about 8000 K when passing through the plasma. When they enter the normal viewing zone (15–20 mm above the coil), the decomposition products

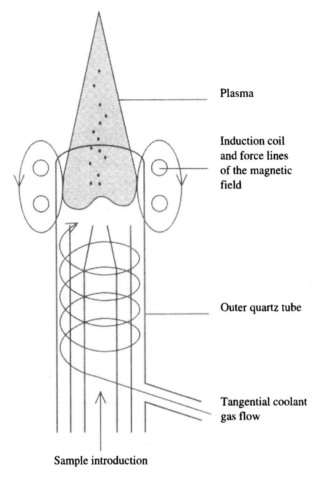

Plasma

Induction coil
and force lines
of the magnetic
field

Outer quartz tube

Tangential coolant
gas flow

Sample introduction

Figure 117 *Schematic diagram of an ICP torch*

of the sample have been delayed for 2 ms in the plasma at 8000–5500 K.
The delay-time and the temperature which particles experience in the
ICP are about twice those of a dinitrogen oxide-acetylene flame. In
contrast to various combustion flames, free atoms and ions are in a
chemically inert atmosphere in the ICP, and this is one reason why their
lifetime in the plasma is longer than in the flame.

5.2.1.1 Types of Inductively Coupled Plasmas. Inductively coupled
plasmas are divided into two main groups: (i) high-power, nitrogen-
argon ICPs, and (ii) low- or medium-power argon ICPs. High-power
ICPs are also called 'nitrogen-cooled ICPs' since these plasmas
are operated with a fairly large torch and nitrogen as the outer gas

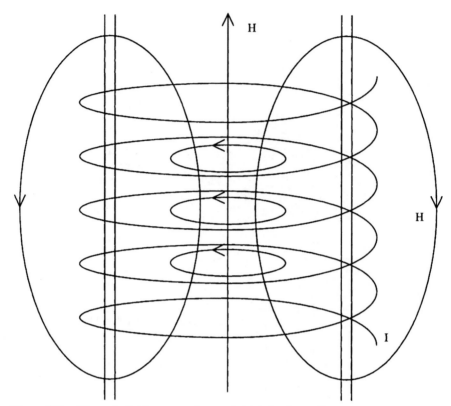

Figure 118   *Magnetic field generated around the induction coil*

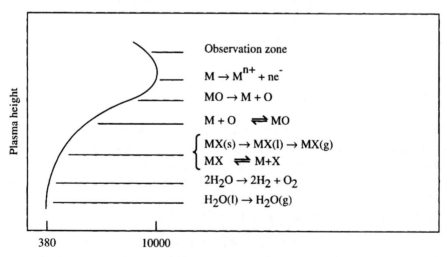

Figure 119   *Reactions of the sample (MX) in an ICP*
(Adapted from: J.-O. Burman and K. Boström, Kemisk Tidskrift, 1980, **92** (8), 28)

(coolant) and argon as the intermediate and carrier gas. Plasma gases other than argon (e.g. N_2) offer some advantages compared to argon plasmas. Thermal conductivities of molecular gases are often higher to that of argon and hence better energy transfer to sample aerosol is obtained. This in turn allows higher solvent loading of the plasma and better analytical performance in the analysis of refractory compounds. On the other hand, some spectral interferences are introduced. These originate from the emissions of CN and NO, when N_2 is used as a plasma gas. In practice, most of the commercial instruments use pure argon as a plasma gas.

The torches for low- and medium-power plasmas are narrower and generally only two argon flows (outer and carrier) are used when aqueous sample solutions containing only inorganic components are to be analysed. When the sample also contains organic material, three argon flows are required. In Table 30, the ranges of the operating parameters used with the conventional high-power and low- or medium-power plasmas are listed. Both types of ICP are very stable and when analysing the same sample repeatedly the relative standard deviation is about 0.4%.

5.2.1.2 Radio-frequency Generators. Radio-frequency (RF) generators are required to supply power to the ICPs. These are oscillators which generate an alternating current at a desired frequency. The basic circuit (a 'tuned circuit' or 'tank circuit') consists of a capacitor and an inductor (coil) in parallel. When the capacitor is discharged through the inductor, the collapse of the magnetic field causes charge build-up on the capacitor. This charge is opposite in polarity to the original charge. The oscillatory process would continue indefinitely, if there was no resistance in the circuit. However, the oscillation will gradually decrease unless enough electrical energy is transferred into the tuned circuit. The oscillation may be sustained by connecting the tuned circuit to the

Table 30 *Ranges of operating parameters for conventional high-power and low- or medium-power ICPs*

Parameter	Low- or medium-power ICPs	High-power ICPs
Frequency (MHz)	27–50	7–27
Power (kW)	1–2	3–7
Outer gas flow(l min^{-1})	15–20 (Ar)	20–50 (N_2)
Intermediate gas flow (l min^{-1})	0–1 (Ar)	10–20 (Ar)
Carrier gas flow(l min^{-1})	0.5–1 (Ar)	1–2 (Ar)
Observation height above coil (mm)	12–18	4–10
Sample uptake rate(ml min^{-1})	1–2	1–2

anode of a vacuum tube. A small portion of the amplified voltage is fed back in the correct phase to the grid of the vacuum tube. Thus, enough energy will be fed into the tuned circuit to overcome the resistance losses.

The various oscillators differ mainly in the feedback process used. The two main types of oscillator used in plasma spectrometry are 'free running' and 'crystal controlled'.

In the free running oscillators, the basic frequency of oscillation is fixed by the values of the components in the tuned circuit. These are in turn modified by any changes in the plasma impedance and in the coupling of the plasma to the induction coil. A piezoelectric crystal is used in the crystal controlled oscillators to control the feedback and to maintain constant frequency. In practice, there are no differences in the operation of these two oscillator types, and they both produce good results.

A number of commercial instruments use generators with nominal powers of 2 kW and frequency of 27.12 MHz. Higher frequency of oscillation may lead to lower excitation and ionization temperatures, lower electron densities, and lower continuum background, together with improved background stability. Improved stability together with lower background emission gives improved detection limits which can be clearly seen as the frequency is increased from 27 to 40 MHz. Nowadays a frequency of 40.68 MHz is commonly used.

5.2.2 Direct Current Plasmas

The stabilized DC arc plasmas form a heterogeneous group of plasma excitation sources. They differ greatly in construction, with variations from the conventional DC arc to the 'free-burning' flame-like plasmas. Their spectroscopic properties are also different.

A feature common to all these DC plasma sources is the DC arc discharge. The discharge is stabilized in various ways or transferred away from the arc column to produce a flame-like appearance (plasmajet). DC plasmas are divided into two groups: (i) current carrying DC plasmas, and (ii) current free DC plasmas or plasmajets.

(i) Current carrying DC plasmas. These may be further divided into gas stabilized types and disc or wall stabilized types. In these plasmas the DC arc discharge is cooled in the peripheral regions by a gas stream or by narrow walls. In this manner the electrical conductivity goes down and the current is concentrated in the inner region of the plasma. The current density increases which also causes an increase in temperature. Further, the high current density causes a magnetohydrodynamic pinch

using the self-induced magnetic field which constricts the discharge even more. In this way the DC arc becomes stable and acquires a higher energy density. The wandering and flickering of the plasma column found with conventional arc excitation are specifically minimized. Extremely high temperatures and currents have been reached in this way by DCPs.

At temperatures above 5000 K, the plasma is very viscous, and hence the effective sample introduction into the plasma is difficult. The current carrying plasmas, especially, strongly resist the introduction of aerosol particles because of the cooling effect of the impinging flow. However, the mixing of plasma and sample aerosol with current free plasma jets is easier.

(ii) Plasmajets. In the current free plasmajets, the plasma is carried out from the arc column by a gas stream, and a 'free-burning plasma flame' (in which current is not flowing) is formed.

Two research groups independently introduced, in 1959, a DC plasmajet for solution spectroscopy. These plasma sources consisted of two ring-shaped carbon or graphite electrodes one on top of the other (an anode above and a cathode below). A DC plasma formed between these electrodes in a chamber with an arc current from 15 to 25 A. Argon or helium was used as the stabilizing gas and it was introduced tangentially into the space between the two electrodes. The plasma was transferred through the cathode ring by the vortex coolant flow. A plasmajet (the actual radiation source) appeared above the cathode and reached a height of about 1 to 2 cm. With this construction it was also a great problem to introduce the sample solution into the plasma. The stability of the plasma was later improved by an external cathode (a tungsten rod) placed above the two ring electrodes. The plasma was then formed between the ring anode and the tungsten rod cathode.

However, these types of plasmajets were prone to several disadvantages. The arcs were unstable and wandered about on the electrode surfaces. The electrodes, eroded by the arc, contributed to spectral emission as did the coolant and carrier gases. The mixing of the sample aerosol was not satisfactory.

Valenta and Schrenk, in 1970, introduced a plasmajet which consisted of two separate chambers for the anode and the cathode. After ignition of the arc, these electrodes formed an angle of 30° to each other. The chambers were aluminium tubes (10 mm in diameter). At the upper ends were electrically isolated water-cooled control orifices, and at the other ends were water-cooled electrode supports. The aerosol sample was carried into the anode chamber with argon. Only argon flowed into the cathode chamber.

A DC arc was ignited by putting the cathode rod in contact with the anode and then withdrawing the cathode to its own position. The two plasma tongues united and formed one plasmajet in which the sample was unsymmetrically distributed. However, sample desolvation was necessary since a directly injected aerosol quenched the discharge and wet aerosol from an indirect nebulizer caused poor sensitivity.

The modern plasmajets are three electrode devices (Figure 120). These DC plasmas are commercially available and are being used in many laboratories. The plasma is sustained by two DC arc-paths which

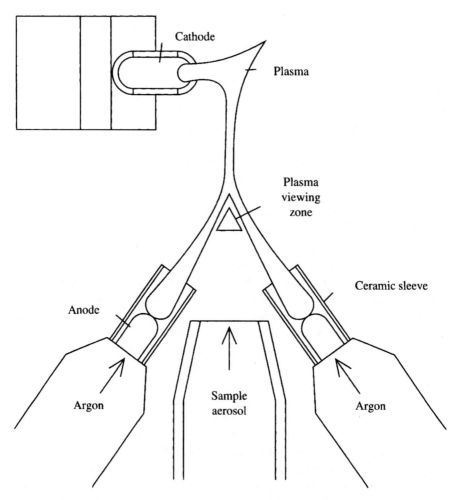

Figure 120 *Schematic diagram of a three-electrode DC plasmajet*
(Fison Instruments)

discharge from the cathode to the two anodes with a shape like an inverted letter 'Y'. The temperature in the middle of the plasmajet may rise up to 10000 K. At this temperature argon is ionized and excited, which results in a high background emission. In order to avoid the interference caused by argon, a three electrode system is used consisting of two graphite anodes and a tungsten rod cathode, all of which are water cooled. Argon flows tangentially through ceramic sleeves co-axially arranged around the electrodes. An aerosol sample (carried by argon) enters the plasma upwards through a wide sample introduction tube. The viewing zone is a triangular arc with a fairly constant temperature of about 5000 to 6000 K between the anodic jets.

This three electrode construction gives a very stable plasma source, and relative standard deviations of 0.5 to 1% may be obtained. The power requirement is only 700 W and the argon consumption is moderate at a rate of 7 l min^{-1}. In practice, the electrodes and sleeves must be changed every few hours. In contrast to the argon ICPs, fewer strong ion lines are observed with the three-electrode DCP.

5.2.3 Microwave Plasmas

Microwave plasmas are divided into two groups: (i) microwave induced plasmas, MIPs; and (ii) capacitively coupled plasmas, CMPs. The construc tion of these two plasma types is very different, and they also differ significantly in their efficiency and analytical possibilities. However, the working frequency of both plasma types is 2450 MHz with few exceptions.

Microwave generators at powers from 100 W to a few kilowatts are driven by magnetrons (Figure 121). The heated cathode emits electrons which flow outwards to an annular anode containing cavity resonators. A very strong stationary magnetic field flows perpendicular to the plane of paper. The electrons are accelerated by the circling electromagnetic field according to phase and they then return back to the cathode. The returned electrons knock out additional electrons. They also contribute to the heating of the electrode, or they are retarded and transfer a part of their energy to the field. The microwaves are led out from the magnetron by co-axial or cavity waveguides.

When the microwave energy is coupled with a gas stream, a microwave or cavity resonator is required. It is a closed metal tube of rectangular a circular cross-section, and its inner dimensions must be such that standing waves can arise.

The adjustment of the impedance is especially important when aqueous solutions are injected into a low-power MIP. Often this is not

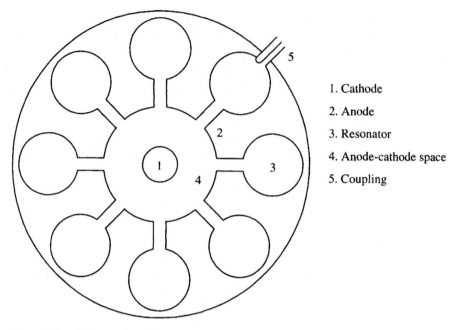

1. Cathode

2. Anode

3. Resonator

4. Anode-cathode space

5. Coupling

Figure 121 *Cross-section of a magnetron*

possible because water and organic compounds may extinguish the plasma.

(i) Microwave induced plasmas. A schematic representation of the structure of microwave induced plasma source is given in Figure 122. The main components of MIPs are the microwave generator, the microwave resonator with wave guide, the quartz capillary, and an injection system.

Normally, argon is used as the plasma gas, but helium and nitrogen have also been used. The gas flow rate lies between 50 and 1000 ml min^{-1}. Ignition of the plasma takes place at the lowest possible power since the magnetron can be damaged if the reflected power is too high. There are two methods of igniting the plasma: *(a)* a Tesla coil, or *(b)* by a spark flash.

The properties of the MIPs are similar to those of CMPs but, due to the low power of the MIPs, the CMPs are superior. Electron densities vary between 10^{12} and 10^{15} cm^{-3}, and they depend on the nature of the plasma gas, the input power, and the pressure. The temperature of an argon plasma at 100 to 1.6 kPa lies normally between 5000 and 6000 K, and 4000 and 4300 K, respectively. At low temperatures the plasma is not in thermodynamic equilibrium. However, at 100 kPa values up to 8500 K have been reported.

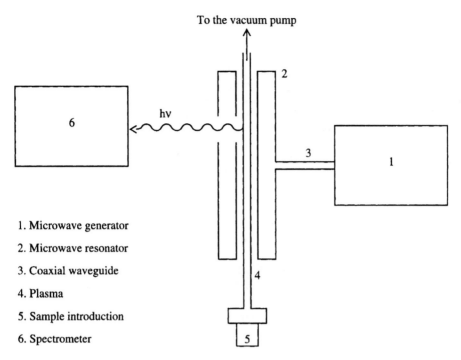

To the vacuum pump

hν

6

2

3

1

4

5

1. Microwave generator

2. Microwave resonator

3. Coaxial waveguide

4. Plasma

5. Sample introduction

6. Spectrometer

Figure 122 *Schematic diagram of an microwave induced plasma source*

The intensity of the emission lines produced by a MIP is affected by the following factors: *(a)* power absorbed from the plasma; *(b)* gas and electron temperature; *(c)* nature, pressure, and flow rate of the plasma gas; *(d)* sample vaporization; *(e)* sample size and composition; *(f)* observed zone of the plasma.

Although the electron temperatures in the MIP are high (about 5000 K), the gas temperatures are relatively low (from about 1000 to 2500 K). Thus, there is too little kinetic energy for dissociation of thermally stable compounds. The vaporization of the sample depends on the sample introduction system used. MIPs are used as detectors for gas chromatography.

(ii) Capacitively coupled microwave plasmas. A capacitively coupled microwave plasma source (CMP) was developed by Cobine and Wilbur in 1951.

The power of the microwave generators is between 0.5 and 3 kW considerably higher than those of the microwave induced plasmas. The frequency is the same (2450 MHz).

The principle of a CMP is shown in Figure 123. A magnetron and power supply are placed in the microwave generator. A waveguide

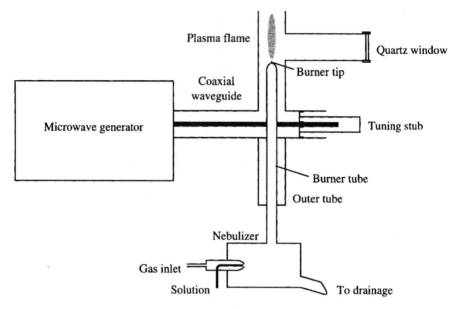

Figure 123 *Schematic diagram of a CMP source*

system (a cavity or a co-axial waveguide) transfers the ultra high frequency energy into the inner plasma tube. The plasma is ignited by a spark discharge which produces the first electrons and ions.

CMPs are well suited for direct solution analysis since the high power and larger amounts of the solvent do not affect the plasma very much. The higher power consumption causes the gas and excitation temperatures to approach each other, but the excitation temperatures still lie above the gas temperatures and the plasma is not in thermodynamic equilibrium. The maximum gas temperatures lie between 4500 and 5700 K, while the excitation temperatures vary between 4900 and 8150 K. The temperatures increase with increasing ionization potential of the plasma gas. Excitation temperatures for MIPs and CMPs are nearly the same, whereas the gas temperatures for CMPs are considerably higher. For argon and helium the gas temperatures remain much lower than for nitrogen.

The main source of interference is the presence of elements with low ionization potential. They may alter the intensities of the emission lines by several orders of magnitude. The introduction of the sample into the plasma is also an important factor. The best results have been obtained by introducing the sample through the central core of the plasma as in the ICPs.

5.3 SAMPLE INTRODUCTION

Normally samples are introduced as solutions into the plasma, but the direct introduction of solids and gases is also possible. Hydride and cold vapour methods are also applied to plasma atomic emission spectrometry. In addition, plasmas can be used as detectors for gas and liquid chromatographs (Chapter 9).

5.3.1 Solutions

The sample solutions are made into aerosols by a nebulizer system and transported into the plasma with a carrier gas. Pneumatic, ultrasonic, or grid nebulizers are generally used for nebulizing the sample. Water or organic solvents can be removed from the aerosol by a desolvation unit if the plasma source cannot take large quantities without interference.

5.3.1.1 Pneumatic Nebulization.   Pneumatic nebulizers are of two basic types: concentric nebulizer and cross-flow nebulizer. The viscosity and surface tension of the sample solution affect the carrier gas flow rate and the liquid uptake. In order to ensure steady sample uptake, peristaltic pumps are used for sample introduction.

Three different types of aerosol generation have been recognized: dropwise, stringwise, and filmwise breakup of the liquid. The breakup pattern changes as the velocity difference (relative velocity, equation 46) between the gas and liquid changes, almost independently of the nebulized liquid. When the relative velocity change increases from about 5 to $50 \ m \ s^{-1}$, the breakup pattern changes from dropwise to stringwise to filmwise. For the ICP nebulizer systems commonly used, liquids will break up in a filmwise manner (the relative velocity being more than $50 \ m \ s^{-1}$).

The most widely used nebulizers in ICP spectrometry are the glass concentric nebulizers (Meinhard nebulizers, Figure 28), which are supplied in varying configurations. The input pressure of the Meinhard nebulizer ranges from 70 to 350 kPa, and the sample uptake rate ranges from 0.5 to 4 ml min^{-1}, when the flow rate of argon is 1 l min^{-1}. At the lower argon flow rates (0.5–0.8 l min^{-1}), a liquid uptake rate of less than 0.5 ml min^{-1} can be obtained. The low argon flow rate is suitable for the small ICP torch (Fassel torch), while it is not enough for the large ICP torch (Greenfield torch) which requires an auxiliary injection gas. The Meinhard nebulizer has a good long-term stability, but it suffers from the blockages caused by particles in sample solutions having too high a solute concentration because of the very small annular gap (10 to 35 μm in diameter).

The first cross-flow nebulizer for ICP spectrometry was designed by Kniseley *et al.* in 1974. In this nebulizer two adjustable capillaries (a vertical tube for liquid uptake and the other one for the injector gas stream) were set at right angles to each other in a poly (tetrafluoroethylene) (PTFE) body. The gas flow rate of argon is about 1 l min^{-1} and the liquid uptake rate is approximately 3 ml min^{-1}. Maximum performance is obtained by adjusting the positions of the capillary tips. However, the long-term stability of these cross-flow nebulizers with adjustable capillaries is not good and, hence, fixed cross-flow nebulizers with nonadjustable capillaries and an impact bead are often used. Cross-flow nebulizers are less prone to blockage at the tip when introducing solutions having high solute concentrations.

The relative movement of the adjustable glass needles was blamed for causing the poor long-term stability of the cross-flow nebulizers. In addition, the vibration of the inner capillary in the concentric nebulizer results in lower precision than would otherwise be obtained. In order to overcome these problems Anderson, Kaiser, and Meddings, in 1981, introduced a cross-flow nebulizer with nonadjustable thick-walled glass capillary tubes (the MAK nebulizer). The MAK nebulizer operates at pressures (approximately 1400 kPa) much higher than commonly used with the Meinhard nebulizer. The argon flow rate is about 500 ml min^{-1}. MAK nebulizers exhibit good long-term stability, good precision (RSD less than 0.5%), and good resistance to blockage at the capillary tip. For example, sodium chloride solutions with concentrations up to 30% can be introduced without problems.

The Babington nebulizer is a variation of the cross-flow nebulizer. Babington nebulizers are not prone to blocking by particles or concentrated solutions, because the liquid emerges through a relatively large slot or hole in the surface. Several variations of this nebulizer have been introduced. GMK Babington nebulizers are commercially available. The argon flow rate ranges from 0.8 to 1.2 l min^{-1} and the operating pressure is 280 kPa. The performances of the GMK nebulizer are better than or equal to those of the cross-flow and concentric nebulizers with respect to detection limits, precision, memory effects, sample analysis time, particle handling capacity, salt effects, and acid effects. Also saturated solutions and those containing particles can be introduced without any nebulizer blockage.

5.3.1.2 Ultrasonic Nebulizers. Ultrasound can be used to break up a liquid mass into smaller particles. In the ultrasonic nebulizer, an ultrasonic generator drives a piezoelectric crystal at a frequency between 200 kHz and 10 MHz. The surface of the liquid sample will breakdown into an aerosol when the longitudinal wave propagates from the surface of the crystal toward the liquid-air interface. The wavelength of the surface wave is given by the following equation:

$$\lambda = (8\pi\sigma/\rho f^2)^{1/3} \tag{80}$$

where λ is the wavelength, σ is the surface tension, ρ is the liquid density, and f is the ultrasonic frequency. The average droplet diameter (D) may be obtained by the following equation:

$$D = 0.34\lambda \tag{81}$$

There are two types of ultrasonic nebulizer (the liquid coupled and vertical crystal systems). The principles of these nebulizers are described in Section 3.2.2.

The production of aerosol is very large and independent of the gas flow for both types of ultrasonic nebulizer. More analyte can be transported to the plasma at slower flow rates of the carrier gas than in the case of the pneumatic nebulizer. Residence time of the analyte in the plasma is longer and subsequently more emitting species are obtained. In general, detection limits are about an order of magnitude better with ultrasonic nebulizers compared with pneumatic nebulizers. On the other hand, effects of background shifts and spectral coincidences will also be enhanced to a similar extent. In both nebulization systems problems caused by the blockage of the capillary orifice increases when solutions having high solute concentrations are analysed. This effect may be overcome by the addition of a laminar gas flow surrounding the aerosol stream. Interferences which are dependent on the particle size are reduced because the droplets produced by a 1 MHz ultrasonic nebulizer from aqueous solutions are smaller than 0.4 μm in diameter.

5.3.1.3 Grid Nebulizer. The grid nebulizer design features a fine-mesh grid of inert material positioned in front of an argon stream (Figure 124). The sample solution is pumped across a Pt screen, wetting it completely. A high-velocity argon stream blows droplets of the solution out of the tiny orifices of the screen. The nebulizer is very efficient, converting a high proportion of solution into a very fine aerosol. A second Pt screen acts to further break down particle sizes and space between the two screens damps pulsations caused by the peristaltic pump.

The grid nebulizer is suitable for analysis of highly concentrated solutions. The short-term stabilities of the grid and fixed cross-flow nebulizers are comparable for solutions such as 2% HNO_3, but the grid nebulizer exhibits better RSD values for aqueous solutions which

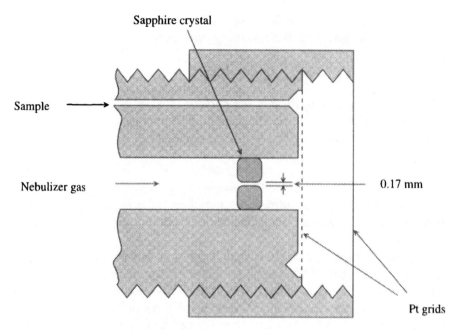

Figure 124 *Hildebrand grid nebulizer*
(Unicam Analytical systems Ltd.)

contain dissolved solids in high concentrations. The grid nebulizer exhibits practically no clogging problems or memory effects.

5.3.1.4 Spray Chambers and Desolvation System. A nebulizer must produce droplets less than 10 µm in diameter in order to achieve a high aerosol transport efficiency (the percentage of the mass of nebulized solution that reaches the plasma), and rapid desolvation, volatilization, and atomization of the aerosol droplets. Pneumatic nebulizers, especially, produce highly polydispersive aerosols with droplets up to 100 µm in diameter and these large droplets must be removed by a spray chamber (Figure 125). The spray chamber may be cooled to 3–4 °C to reduce solvent overloading of the plasma.

A barrel-type (Scott) spray chamber removes the large droplets by turbulent deposition on the inner walls of the chamber or by gravitational action. Small pressure changes in the chamber may affect the plasma. In addition, the fall of the drops in a poorly designed drain may cause spikes of noise on the recorded signal. Liquid build-up in the chamber causes changes in pressure, which in turn change the velocity and cause drift in the signal. The shape, size, and position of baffles or other internal parts of the spray chamber affect the rise and decay of the spectral signal. The carrier gas velocity and the analyte element also affect the signal.

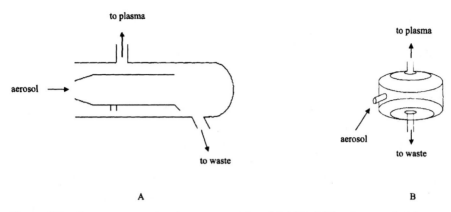

Figure 125 *Common spray chambers types employed in ICP-AES technique: double pass spray chamber (A) and cyclonic spray chamber (B)*

The nebulization rates achieved with the spray and nebulizer systems used in ICP spectrometry are much slower than those used in flame atomic absorption. The transport efficiency of the sample introduction systems is less than 3% in ICP spectrometry, whereas that in flame atomic absorption is about 5% or less. After introduction of concentrated solution into the plasma the emission signal returns to baseline quite slowly. One has to wait generally more than 40 seconds until the next measurement can be made to avoid systematic errors caused by carry-over effects. The wash out times encountered in ICP-AES technique are thus much longer than in flame atomic absorption spectrometry. However, with the use of cyclonic spray chambers somewhat lower wash out times are obtained (Figure 125).

One disadvantage of high-efficiency nebulizers (eg. ultrasonic nebulizer) is that they convey large amounts of solvent to the plasma. Solvent loading will cool down the plasma and deteriorate its excitation efficiency. Small, low-power ICPs may even be extinguished. Solvent loading is a problem especially when organic solvents are introduced. Desolvation may be achieved by heating the aerosol sufficiently to evaporate the solvent from the aerosol. A heated spray chamber or a separately heated tube may be used for desolvation. The resulting vapour and dried particles are then passed through a condenser which removes most of the solvent in the vapour. However, some of the dried particles are also lost. The liquid solvent is removed through the drain tube and the dried particles are transported to the plasma. Sample aerosol may be further dried with the aid of cryogenic desolvation system having temperatures well below centigrade zero. The desolvation process may cause interferences if there is a difference between the volatilities of the analyte and the matrix.

Currently also various membrane desolvation systems are used. Hygroscopic nafion membranes are suitable for polar molecules, and microporous PTFE membranes can be used for many organic solvents, like hexane, methanol and dichloromethane.

5.3.1.5 Nebulizers for Low Sample Volumes. Usually the sample introduction rate in ICP-AES (and in ICP-MS) is about $1-2$ ml min^{-1}, when ordinary nebulizer-spray chamber systems are used. This might be a problem when the sample volume available is limited, or when hazardous or expensive samples are analyzed. Therefore special microconcentric nebulizers (MCNs) are constructed where the sample flow rates might be well below 100μl min^{-1}. When these devices are equipped with desolvation systems, a very low solvent loading of plasma is obtained.

Recently there has been an increasing interest to introduce samples into the plasma without using a spray chamber. In this approach (Direct Injection High Efficiency Nebulizer, DIHEN) a microconcentric nebulizer is inserted inside the plasma torch instead of sample injector tube (Figure 126). This results in very high sample introduction

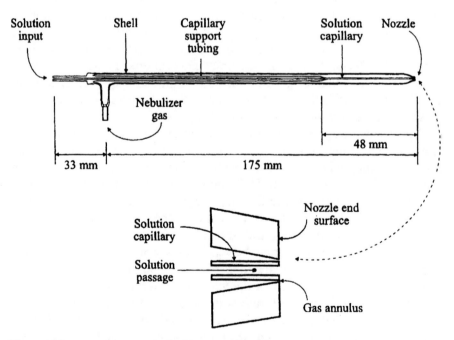

Figure 126   *Direct injection high efficiency nebulizer. The capillary internal diameter is 82 μm and gas annulus area 0.0099 mm²*
(Adapted from: J.A. McLean, H. Zhang and A. Montaser, Anal. Chem., 1998, **70**, 1012)

efficiency (close to 100%). The sample uptake rate is very low, typically $10\,\mu l\,min^{-1}$. Therefore syringe pumps are used for transporting the sample to the nebuliser. Additionally, an injection valve can be used, allowing a constant sample flow to the nebuliser for a definite period of time, depending on the flow rate and the volume injected.

The use of DIHEN removes/reduces some drawbacks associated with ordinary spray chamber-nebuliser systems. Very low internal volume of DIHEN results in rapid response times and memory effects are also reduced. In real samples analytes may be present in many chemical forms that may show variations in chemical and physical properties (such as evaporation rate in a spray chamber). This in turn may result in different sensitivity for different species and systematic errors when total concentrations are measured. This behaviour may be diminished (but not completely removed) by the use of DIHEN. A drawback of DIHEN is an increase solvent load of the plasma (especially when organic solvents are used) that may result in deterioration of the plasma analytical properties.

5.3.2 Solid Samples

Many samples occur naturally in solid form and a number of advantages can be gained if solid samples are introduced directly into the plasma source. Pretreatment or dissolution steps are not required, and contamination from reagents is minimized. Dilution errors are also eliminated and sample losses are avoided. Sample preparation is the most time-consuming step in conventional plasma spectrometry when samples are introduced in solution form into the plasma. The direct analysis of solid samples reduces reagent and manpower costs. Improved absolute detection limits may be obtained, analysis of microsamples or localized segments of samples are possible, and the sample vaporization, atomization, and excitation steps may be separated and optimized.

Various techniques have been employed to introduce solid samples directly into plasma. These techniques may be divided into two groups: (1) direct insertion of samples into the plasma, and (2) methods that convert solid samples into an aerosol or a vapour, which is then transported into the plasma. The latter techniques include electrical spark and arc ablation, electrothermal vaporization, and laser ablation.

5.3.2.1 Direct Insertion of Samples. In direct insertion the sample is introduced on a probe directly into the plasma. The sample is usually ground into a powder or segmented into sufficiently small pieces. Samples often must be dried before weighing because many samples contain variable amounts of water. The probe for sample insertion is made of graphite, tantalum, or tungsten. The insertion process is carried out manually or automatically in one or a series of specified steps. Signals are measured as time-dependent intensities because differential volatilization occurs. The probe material and geometry, the analyte, composition of the sample matrix, and the introduction sequence affect the time dependence.

In the first reported direct sample insertion device, the ICP torch was modified by replacement of the inner tube with a quartz tube that acted partially as a guide for the sample probe (Figure 127). A graphite sample probe is placed on the end of a solid quartz rod that can be inserted into the inner quartz tube. The bottom end of the quartz rod is attached to a sliding platform. The graphite electrode probe is moved into the plasma by moving the platform manually into the vertical position. Introduction of air into the plasma must be prevented.

By using automatic position devices, such as stepper motors, DC motors, electrically operated car aerial motors, a combination of pneumatic actuators and motor-driven screw assemblies, and pneumatic elevators, the positions of the sample probe (intermediate and final positions) and duration at each position as well as rate of insertion are easily controlled. These systems are under computer control and relatively simple software is needed to achieve digital control of sample positions and duration at each position. Positioning of the sample probe at successively closer locations to the plasma enables sample pretreatment and atomization steps, analogous to drying, ashing, and atomization stages used in GF-AAS. Instrumentation to automate changing of sample probes has also been described.

Many of the operating parameters normally optimized for ICP-AES with solution nebulization, are also important in direct sample analyses. These include observation height, RF power, gas flow rates, integration time, and wavelength. In direct sample insertion, sample insertion heights for drying, ashing, and vaporization, rate and duration of insertion, and sample size must also be optimized.

Determinations can be made in the range of micrograms per gram, with the possibility for determination at submicrogram per gram. Precision (RSD-values) ranges from 1 to 25%, being typically from 7 to 10%. The relatively poor precision is primarily caused by the sample-to-sample heterogeneity, vaporization irreproducibility, and imprecise sample insertion. Both precision and detection limits can be improved by addition of volatilization enhancement reagents such as sodium fluoride. Calibration graphs are generally linear over at least three orders of magnitude. Successful calibrations have been achieved by

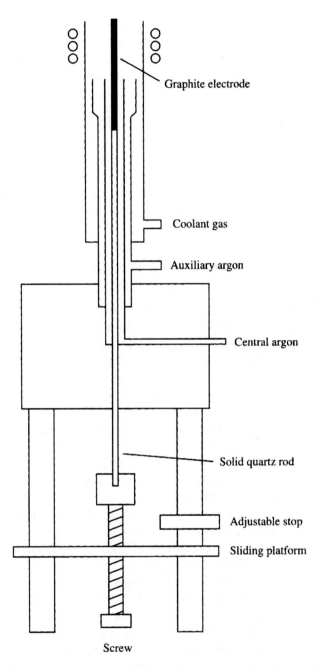

Figure 127 *Schematic diagram of a direct sample insertion device for ICP-AES*
(Adapted from: E. D. Salin and G. Horlick, Anal. Chem., 1979, **51**, 2284)

NIST standard reference materials (especially, coal, botanical and nickel-based alloy samples) and Spex G standards.

The primary interference in the direct insertion technique is the offset intensity due to background shifts during the sample volatilization. These shifts are generally small and they can be corrected by normal background correction methods, but the transient nature of the signals complicates the procedure. Other limitations of the direct insertion technique are related to the probe material, as in DC arc spectrometry. Shortcomings of graphite such as reactivity, carbide formation, and occasional memory effects may be reduced by using pyrolytically coated graphite, metal cups, volatilization enhancement reagents (analogous to matrix modifiers in GF-AAS), hydrogen as an oxygen scavenger, and preburn and postburn cycles.

5.3.2.2   Methods that Convert Solid Samples into an Aerosol or Vapour.   These methods include (i) electrothermal vaporization, (ii) arc and spark ablation, and (iii) laser ablation.

(i) Electrothermal vaporization (ETV): Electrothermal vaporization techniques similar to those utilized in AAS are used in plasma AES. The volatilization of the sample is performed by resistive heating of a sample-supporting device. These devices include filaments, boats, ribbons, rods, and tubes constructed of graphite or a refractory element. Most of the ETV systems utilize a power supply for the vaporizer that can be sequenced through several heating steps like those in GF-AAS. By the use of proper heating programmes, the solvent, matrix components, and analytes can be temporally separated from each other.

The sample is placed on the surface of the vaporizer in solid or in liquid form. At the vaporization stage a dry, highly dispersed aerosol is produced. The aerosol is subsequently swept into the plasma by the carrier gas (usually argon) and a transient emission signal is obtained.

A number of different electrothermal vaporization devices have been used. These vaporizers are contained within argon-filled enclosures. In the graphite rod vaporizer, the sample rod is positioned between two water-cooled terminals of the power supply, and the assembly is surrounded by a cylindrical glass manifold. The sample is inserted into the graphite rod through a sample delivery port located on the conical top of the manifold. The top of the manifold also contains another port which permits the vaporized sample to be swept into the plasma by the carrier gas. The graphite rod is heated by a programmable, high-current power supply.

In one ETV device the electrothermal vaporization chamber is connected directly to the ICP torch assembly. In this design the distance that the sample must travel from the surface of the vaporizer to the plasma is kept minimal (about 30 cm) and the vaporization chamber is small (about 50 cm³) in order to prevent diffusion of the analyte peak and to minimize loss of the sample to the walls of the delivery system. The sample is inserted into a tantalum boat, mounted on brass supports through which the power is applied.

Another ETV system is a modified version of an AAS atomization chamber. The sample is placed on a pyrolytically coated graphite microboat supported by a carbon rod and its mounting assembly. The vaporization heating cycle is controlled by a programmable power supply. The vaporizer assembly is connected to the plasma by a PTFE tube (12 cm in length).

ETV-ICP is an extremely sensitive technique due to highly efficient sample utilization. Detection limits for aqueous solutions are generally 10 to 100 times better than those obtained using solution nebulization. For solid samples, detection limits fall to the nanogram-to-picogram range. The sample size typically used is 0.5 mg.

The major limitation of solid sample introduction in ETV-ICP is the requirement for sample homogeneity. Because the maximum sample size is 10 mg, weighing constraints and homogeneity restrictions become critical. Homogeneity can be improved by using micropulverization and sieving in the sample preparation. The expansion of the carrier gas as it passes over the hot vaporizer during the vaporization stage causes downward shifts in the continuum background intensity.

(ii) Arc and spark ablation: Using an electrical discharge to remove particles from the sample surface is one of the most successful methods of introducing solid samples directly into the plasma. The types of discharge may be classified into arcs and sparks. Depending on the conditions employed, the sample material may be vaporized, eroded, or sputtered. Discharges are usually operated in an inert atmosphere such as argon.

The sample must be conductive, naturally or through addition of a conductive material such as powdered copper or graphite. Liquid, solid, or gaseous samples may be analysed, provided that a reproducible discharge can be formed between the sample and electrode. The sample vapour is transported directly or via an intermediate settling chamber into the plasma using an argon flow. The emission signals generated are often time-dependent and affected by the volatility of the analyte, discharge conditions, discharge stability, and composition of the matrix.

The first and one of the simplest electrical discharge ablation systems uses a DC arc source. This aerosol generator consists of an air-cooled hollow anode and a ground cathode that is in contact with the sample. The electrode system is flushed by argon. A discharge between the anode and sample bearing cathode ablates material. The vapour produced is transported by an argon flow through the hollow anode. Relatively long tubes between the aerosol generator and the ICP may be used.

Radiofrequency arc (RF arc) has also been used for removal of particles of material from the sample surface. In this system a modified ICP torch is used, and the sample is in electrical contact with a ground copper base. The base is water-cooled and argon flow is introduced at the base. A stable filament from the plasma forms between the induction region and the sample, when the plasma discharge is initiated. Boron nitride is used as thermal and electrical insulator between the base and the RF arc.

Most commonly, the aerosol sample is generated by the spark discharge. The spark can range from low to high voltage types of varying pulse durations, frequencies, and waveform shapes. Several commercial spark aerosol generators are available (e.g. those supplied earlier under companies Thermo Jarrell-Ash and Fison Instruments). In the Fison system a spark discharge is formed between a tungsten counter electrode and the conducting sample surface. The resulting aerosol is transported by argon directly into the plasma via a connecting tube 1 m in length.

In addition to the instrumental parameters normally optimized in solution nebulization, in arc or spark ablation the aerosol injector gas flow, discharge energy, and time must also be optimized. Detection limits for solid samples extend down to the microgram-per-gram range for the most sensitive lines. Precision for the spark ablation technique (RSD between 0.5 to 5% for most metallurgical samples) is superior to other solid sample introduction techniques used in plasma AES. Calibration graphs for both arc and spark ablation are linear to at least three orders of magnitude, and the sample matrix practically does not affect the analyte signal.

Although there seems to be minimal interference in arc and spark ablation techniques, some limitations exist. For example, introduction of excessive sample quantities into the plasma may give rise to poor precision, curvature of the calibration graph, memory effects, and instability of the plasma. If insufficient arc or spark energy is used, differential distillation of the sample components may occur.

(iii) Laser ablation: Laser ablation (LA) is related to the arc and spark ablation techniques. A focused laser beam is directed onto the sample causing sputtering of material from the sample surface. The vapour and particles released from the surface are transported by an argon stream into the plasma. Analogous to ETV-ICP, data may be acquired in two modes: time resolved spectra and total signal integration.

In addition to the instrumental parameters normally optimized for solution nebulization, the laser ablation technique requires the optimization of the operation of the laser, the volume of the ablation chamber, and the transport distance from the chamber to the plasma. Detection limits reported in the literature indicate that microgram-per-gram determinations are possible. Precision obtained by laser ablation is somewhat degraded when compared to solution nebulization. For a number of different metal samples precisions between 3 and 11% (RSD) are obtained. The linearity of the calibration graphs is typically two or three orders of magnitude.

Due to the nature of the laser beam, laser ablation is best suited for area-localized microanalysis. In practice difficulties are encountered in the determination of trace and ultratrace elements by LA-ICP-AES due to reduced sensitivity and spectral interferences. Therefore LA-ICP-MS technique is superior in many applications. This is discussed in more detail in Section 6.2.1.

5.3.3 Gaseous Samples

Introduction of gaseous samples into the plasma offers several advantages over liquid aerosol introduction. The transport efficiency for introduction of gases approaches 100%, whereas in pneumatic nebulization more than 95% of the sample solution is discarded. When more analyte is transported into the plasma, improved signal-to-noise ratios and detection limits may be obtained.

By far the hydride generation and cold vapour techniques are the most popular techniques to convert analytes to gaseous forms prior to sample introduction. In addition, some inorganic compounds, such as metal chelates, metal carbonyls, and metal halides, are volatile at ambient or slightly elevated temperatures. The formation of these compounds can be used to convert the analyte to a gaseous form and separate it from the sample matrix prior to introduction into the plasma. Molecular gases or gas mixtures can be introduced directly into the plasma. Chloride, bromide and iodide, for example, can be oxidized to elemental form before introduction into the plasma.

5.3.3.1 Hydride Generation Method. Hydride generation and cold vapour techniques are discussed in Section 2.3.2. Initially these techniques were developed for the quantitative determination of the hydride-forming elements and mercury by atomic absorption spectrometry, but nowadays these methods are also widely used also in plasma atomic emission spectrometry. In the hydride generation technique, hydride-forming elements are more efficiently transported to the plasma than by conventional solution nebulization, and the production and excitation of free atoms and ions in the hot plasma is therefore more efficient. Spectral interferences are also reduced when the analyte is separated from the elements in the sample matrix.

Hydride generation systems. Continuous flow, flow-injection and nowadays to a lesser extent batch approaches are used for hydride generation. Commercial hydride generation systems are available for various plasma spectrometers. In the continuous flow systems the sample and reagent solutions are continuously pumped, usually by a multi-channel peristaltic pump, into a mixing chamber (such as a reaction coil) where the acid sample solution and the sodium tetrahydroborate solution are mixed. After a short period to complete the reaction, the gaseous hydrides formed are separated from the liquids, mixed with the argon injector gas, and swept into the plasma. Usually ordinary gas-liquid-separators shown in Figure 29 are used. However, the acidified sample and the reductant solution can be simply mixed in a Y-junction and pumped to the nebulizer. The spray chamber of the plasma AES instrument acts as a gas-liquid separator and hydrides are introduced into the plasma.

As, Bi, Ge, Sb, Se, and Sn have been determined simultaneously using a batch type hydride generation-condensation (HGC) technique. This method involves the rapid mixing of the acid sample solution and the tetrahydroborate solution followed by the condensation of the volatile hydrides formed at low temperature. Liquid nitrogen may be used for the trapping medium. Large amounts of hydrogen gas produced during the reaction are vented. Then the hydrides are evaporated by placing the condensation trap in a hot water bath, and swept by argon into the plasma.

Figure 128 represents the hydride generation system designed for the DCP equipment. The generator assembly of the system consists of a reaction cell, drying tube, hydrogen delay column, flow meter and valves, and a special sample introduction tube. The acid (HCl) sample solution is introduced into the reaction cell, followed by the introduction of a small amount of sodium tetrahydroborate solution, then the cell is kept closed. After the reaction has been completed (about 30 s), the hydride formed is swept into the plasma with an argon stream. The parameters to be optimized are the sample volume, the concentration and volume of the borohydride solution, the flow rate of the carrier gas, and the reaction time. The optimum sample volume is about 10 ml, and the method allows the determination of the hydride-forming elements in the range 1 to 200 $\mu g \, l^{-1}$.

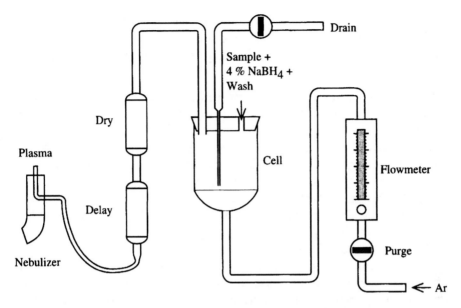

Figure 128 *Hydride generation system designed for a DCP-AES instrument* (Adapted from. T. G. Gilbert, Anal. Lett., 1977, **10**, 599)

Some detection limits reported in the literature for hydride generation techniques are presented in Table 31. With respect to the detection limits published in the literature, hydride-plasma-AES is comparable with hydride-AAS, except for lead. Relative to solution nebulization, detection limits for the hydride generation method are about three

Table 31 *Some detection limits reported in the literature for volatile hydride-forming elements by continuous hydride generation technique and solution nebulization*

Analyte	Detection limit ($\mu g\ l^{-1}$)	
	Hydride generation	*Solution nebulization[a]*
As	0.8[b]	20
Se	0.8[b]	20
Bi	0.8[b]	40
Ge	0.2[c]	200
Pb	1[d]	30
Sb	1.0[b]	20
Sn	0.2[c]	100
Te	1.0[b]	20

[a]K. A. Wolnik, F. L. Fricke, M. H. Hahn, and J. A. Caruso, Anal. Chem. 1981, **53**, 1030.
[b]M. Thompson, B. Pahlavanpour, S. J. Walton, and G. K. Kirkbright, Analyst (London), 1978, **103**, 568.
[c]M. Thompson and B. Pahlavanpour, Anal. Chim. Acta, 1979, **109**, 251.
[d]M. Ikeda, J. Nishibe, S. Hamada, and R. Tujino, Anal. Chim. Acta, 1981, **125**, 109.

orders of magnitude better for many elements. Relative standard deviations reported generally range from 2 to 10%. Depending on the method used, the linear dynamic range varies from 2 to 4 orders of magnitude.

Interferences.   Chemical and spectral interferences are associated with the hydride generation technique. Molecular band interference caused by CO_2, which is generated from contaminant CO_3^{2-} in the $NaBH_4$ solution, is a serious problem at short wavelengths where the most desirable lines for the hydride-forming elements are located. Background from CO_2 interferes specifically with the hydride generation-condensation method. Chromatographic removal of the CO_2 has been used to avoid this problem. For example, CO_2 elutes from the Chromosorb 102 column before the hydrides. Alternately, dynamic correction may be used to compensate for the background emission.

Chemical interferences fall into two categories as in hydride-AAS: (i) Interferences in the liquid phase that prohibit or limit the formation of the volatile hydride, and (ii) Interferences in the gas phase that diminish the analyte available for excitation. Interferences of the first type are caused mainly by the Group 8, 9, 10, and 11 transition elements. The degree and elimination of the interferences are the same as in AAS (Chapter 2, Section 2.3.2).

Gas-phase interferences in atomic absorption cells caused by the mutual interaction of the hydride-forming elements are unlikely in various plasmas because of their high temperatures. However, in DCP severe interference caused by other hydride-forming elements have been noted (Figure 129).

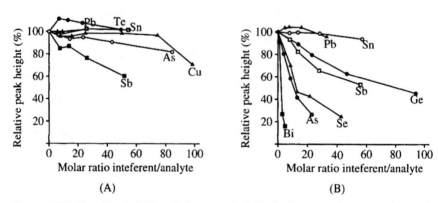

Figure 129   *Interferences caused by various hydride forming elements in the determination of selenium (A) and tellurium (B) by DCP-AES using hydride generation technique*
(Adapted from: H. Häyrynen, L. H. J. Lajunen, and P. Perämäki, At. Spectrosc., 1985, **6**, 88)

5.3.3.2 Cold Vapour Method. Mercury can be determined in plasma AES by reducing it first to elemental mercury and then transporting the mercury vapour into the plasma. The same reduction methods may be used as for AAS. Commercial hydride generation systems can be adopted to the cold vapour method. The detection limit is about 0.02 mg l^{-1}.

5.4 SPECTROMETERS

Each spectrometer for sequential or simultaneous multi-element plasma AES measurements is equipped with a monochromator in order to: (i) have an adequate wavelength selection, and (ii) collect as much light as possible from a selected spectrum area in the radiation source. A monochromator consists of: *(a)* an entrance slit, *(b)* a collimator to produce a parallel beam of radiation, *(c)* a dispersing element (a prism or a grating), *(d)* a focusing element reforming the specific dispersed narrow bands of radiation, and *(e)* one or more exit slits to isolate the desired spectral band or bands.

In plasma atomic emission spectrometry the dispersing element mostly used is a grating. Two types of gratings are employed: (i) 'conventional' gratings, and (ii) echelle gratings. A fore-prism as an order-sorter is an essential feature for use with an echelle grating. The operation of conventional gratings is discussed in Section 2.2.2.

The spectrometers used are adapted either for sequential or simultaneous multi-element measurements. Commonly used grating spectrometers in plasma AES include: (i) spectrometers with the Paschen-Runge mount, (ii) echelle spectrometers, (iii) spectrometers with Ebert and Czerny-Turner mounts, (iv) spectrometers with Seya-Namioka mounts, and (v) double monochromators. Also Fourier transform spectrometers may be used in plasma AES.

5.4.1 Spectrometers with the Paschen-Runge Mount

Spectrometers with the Paschen-Runge mount are the most commonly used direct-reading spectrometers in optical emission spectrometry. The grating and the entrance and exit slits are in fixed positions at the Rowland circle (Figure 130). An exit slit frame with a few mounted slits is shown in Figure 131. High dispersion (reciprocal linear dispersion of 0.3 to 0.4 nm mm^{-1} in the first order) is obtained with a focal length of 1 m and a grating with 2400 grooves mm^{-1}. Normally an entrance slit of 20 μm and exit slit of 40 μm are used.

A Paschen-Runge mount is especially appropriate for multi-element determinations. The slit positions are tuned for various elements and

<m<mXXX

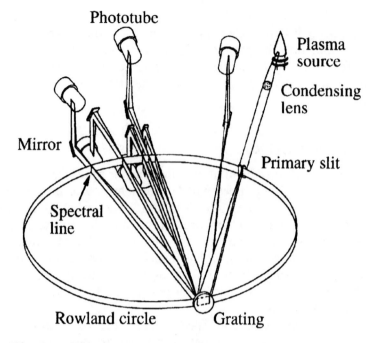

Figure 130 *Optical diagram of the polychromator based on a Paschen-Runge mount*

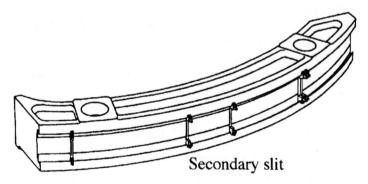

Figure 131 *A secondary slit frame with 5 secondary slits mounted*
(Fison instruments)

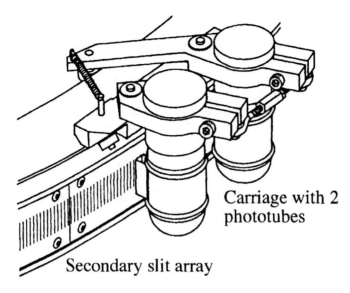

Carriage with 2
phototubes

Secondary slit array

Figure 132 *A sequential spectrometer based on a Paschen-Runge mount*
(Fison Instruments)

separate photomultiplier tubes are placed behind each slit. It may be adapted for sequential measurements by directing the light from each slit using mirrors to a single photomultiplier.

Another design of the Paschen-Runge mount for sequential measurements is shown in Figure 132. Here 255 fixed secondary slits with a regular spacing of 2 mm are mounted in the slit frame instead of the normal individual slits. Two movements are required to set the monochromator at a desired wavelength. First the carriage must be set to the specific exit slit, and then the entrance slit must be accurately moved to the desired spectral line. With an instrument of 1 m focal length a wavelength setting accuracy of 0.0005 nm can be obtained in the second order for a grating with 1080 grooves mm^{-1}. Stray light and interference of overlapping orders may be reduced by placing filters in the primary beam.

5.4.2 Echelle Spectrometers

The echelle grating was invented by Harrison in 1949. An echelle grating is a ruled plane diffraction grating with relatively few rulings (typically 300 grooves mm^{-1} or less). The resolution of a diffraction grating is directly proportional to the spectral order *(m)* and groove density *(N)* (Equation 41). Instead of using large numbers of grooves to achieve very high resolution, the echelle grating increases the blaze angle and spectral order.

The linear dispersion of an echelle grating *(dl/dλ)* is the focal length of the spectrometer *(f)* multiplied by the angular dispersion:

$$dl/d\lambda = 2f\tan\beta/\lambda = mf/d\cos\beta \qquad (82)$$

The echelle grating spectrometer gives high linear dispersion by using a high blaze angle and high spectral orders instead of long focal lengths.

An echelle grating is shown in Figure 133 and appears similar to a normal blazed plane grating except that the 'short side' of the grooves is used. Thus, echelle gratings are used at blaze angles greater than 45°. The use of high spectral orders (usually m = 10 to 1000) at high blaze angles requires very precisely controlled shape of the rulings (at least an order of magnitude over a conventionally ruled grating). The effect of the use of high blaze angle and high orders is shown in Figure 134. In Table 32 and Figure 135 a typical conventional spectrometer and an

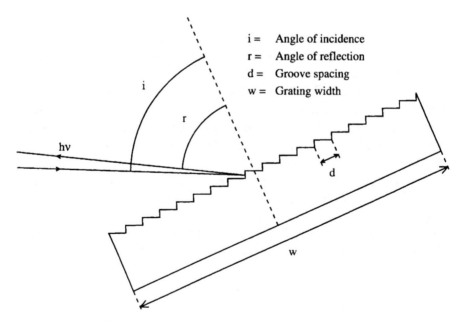

Figure 133   *Echelle grating*
(Adapted from: P. N. Kelihar and C. C. Wohlers, Anal. Chem., 1976, **48**, 333A)

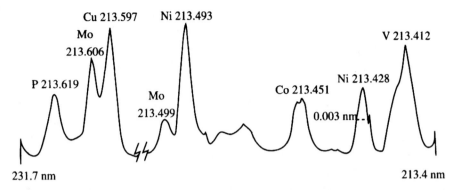

Figure 134 *Wavelength scan which demonstrates the high resolving power of an echelle monochromator (ARL)*

Table 32 *Comparison of typical conventional and echelle spectrometers* (Source: P. N. Keliher and C. C. Wohlers, Anal. Chem., 1976, **48**, 333A)

Feature	Conventional grating	Echelle
Focal length (m)	0.5	0.5
Groove density (grooves/mm)	1200	79
Angle of diffraction	10° 22'	63° 26'
Width of the grating (mm)	52	128
Order at 300 nm	1	75
Resolution at 300 nm	62 400	763 000
Linear dispersion at 300 nm (mm nm^{-1})	0.61	6.65
Reciprocal linear dispersion at 300 nm (nm mm^{-1})	1.6	0.15
f–Number	f/9.8	f/8.8

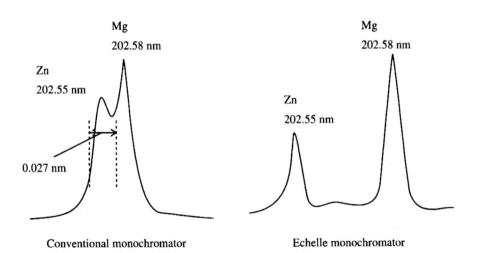

Figure 135 *Comparison of resolution of a conventional and an Echelle monochromator*

echelle spectrometer, both of 0.5 m focal length, are compared. In this case, the echelle spectrometer has better than one order of magnitude higher resolution and dispersion than the conventional spectrometers. However, the overlapping orders in echelle spectrometers must be separated from each other in some way.

Separation of the overlapping orders is done with an auxiliary dispersing element such as a prism or low dispersion grating. If the prism is placed so that its dispersion is perpendicular to that of the echelle a two-dimensional spectral pattern is formed in the focal plane of the system (Figures 136 and 137). The different orders appear as horizontal lines with the lowest order (longest wavelength) at the bottom and the highest order (shortest wavelength) at the top. The prism separates the

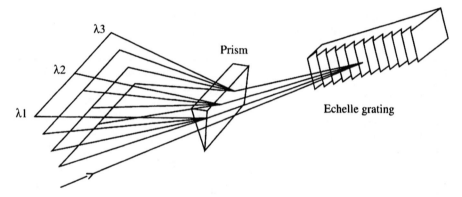

Figure 136   *Echelle monochromator*
(Adapted from: P. N. Kelihar and C. C. Wohlers, Anal. Chem., 1976, **48**, 333A)

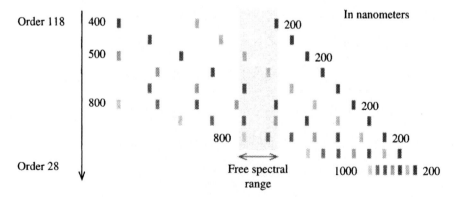

Figure 137   *A two-dimensional spectral pattern obtained by an Echelle spectrometer. The orders are sorted vertically and dispersed horizontally within the orders*

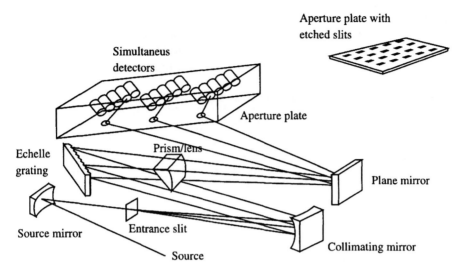

Figure 138 *A sequential echelle spectrometer*
(Unicam Analytical Systems Ltd.)

wavelengths in the vertical direction, while the echelle grating separates them horizontally.

Figure 138 illustrates a sequential echelle spectrometer. Light from the entrance slit is directed by the collimating mirror to the echelle grating, and the diffracted beam is then reflected to the prism lense (order-sorter) and focusing lense. A two-dimensional spectrum is formed in the focal plane of the system. An aperture plate, which contains preset apertures, is mounted at the focal plane. Slits and mirrors are placed at appropriate positions on a mask plate placed above the aperture plate. In this way the light from the selected wavelengths are allowed to reach the corresponding photomultipliers which are mounted in the photomultiplier rack. For three different wavelengths reciprocal linear dispersion and resolution of this commercial echelle spectrometer are listed in Table 33.

The scheme of the light path of another commercial echelle spectrometer is shown in Figure 139 where the echelle grating is mounted in a modified Czerny-Turner configuration. The resulting two-dimensional spectrum is projected onto a screen equipped with tiny slits placed in holes in the exact position for the selected wavelengths. This screen is termed a cassette (Figure 140) and contains the secondary optics, which deflect the light to the photomultipliers. With this arrangement as many as 20 emission lines can be evaluated simultaneously.

The two-dimensional spectral pattern simplifies the use of the echelle as a spectrograph. An optional photographic attachment using 4″ by 5″

Table 33 *Dispersion and bandpass values of a commercial instrument using an echelle grating with 79 grooves mm⁻¹, with a blaze angle of 62° 26', and having the parent line at the 42nd order. Effective aperture of the spectrometer is f/8. The width of the entrance slit is 80 μm, and the width of the exit slit is 40 μm*
(Unicam Analytical Systems Ltd.).

Wavelength (nm)	*Dispersion* (nm mm⁻¹)	*Bandpass* (nm)
200	0.083	0.008
400	0.137	0.016
600	0.205	0.021

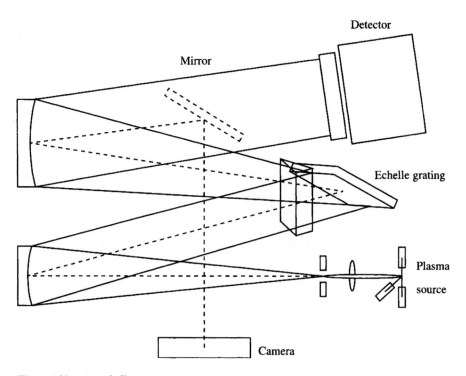

Figure 139 *An echelle spectrometer*
(Fison Instruments)

positive or negative single-frame film allows qualitative analysis. When this attachment is used, a plane mirror is moved into the optical path. The radiation from the focusing mirror is redirected to a focal plane at the front of the instrument. Here a Polaroid® camera is employed to photograph the spectrum of the sample. The resulting 4″ by 5″

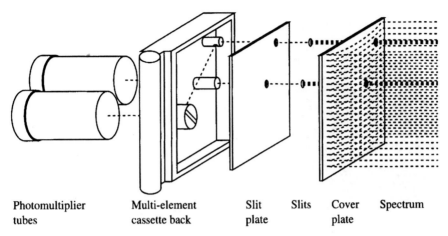

Photomultiplier tubes	Multi-element cassette back	Slit plate	Slits	Cover plate	Spectrum

Figure 140 *An optical cassette for DCP-AES instrument* (Fison Instruments)

photograph is examined using clear plastic overlays on which element locations are indicated (Figure 141). Mercury lines are used as indices for positioning of the overlay.

The introduction of modern charge transfer devices (CTDs) as detector elements for ICP-AES has further improved the usability of echelle spectrometers in simultaneous measurements (Figure 142). These modern detectors contain currently more than one million pixels covering the two-dimensional optical image produced by an echelle spectrometer

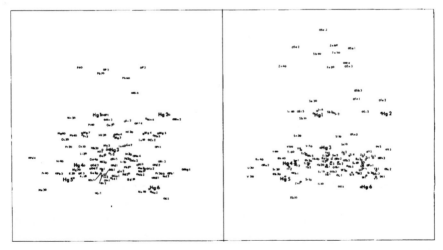

Figure 141 *Plastic overlays for the qualitative analysis by an Echelle spectrometer* (Fison Instruments)

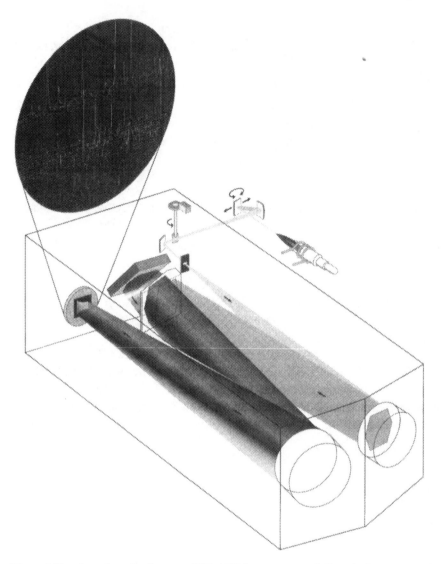

Figure 142 *A modern simultaneous ICP-AES featuring an echelle polychromator and a*
                CCD detector
                (Varian, Inc.)

(echelle polychromator). Tens of thousand emission lines are available
for measurement of over 70 elements in a time less than one minute.
The stability of new instruments is guaranteed by thermostatting the
optical compartment to constant temperature. In addition, the detector
elements are cooled down to −30°C to ensure low noise and low
detection limits.

5.4.3 Spectrometers with the Ebert and Czerny-Turner Mounts

In an Ebert monochromator the entrance slit and exit slit are either side of the grating and a single concave spherical mirror is used as a collimating and focusing element (Figure 143). Wavelength scanning and selection of a specific line are accomplished by pivoting the grating about the monochromator axis.

In a Czerny-Turner mount, two smaller concave mirrors are use instead of a single mirror (Figure 144). The optical characteristics of the Ebert mount and the Czerny-Turner mount are similar. In these mounts changes from one wavelength to another can easily by made by rotating the grating using a computer controlled stepper motor.

5.4.4 Spectrometers with the Seya-Namioka Mounts

The grating is rotated about a vertical axis through the centre of the grating in the Seya-Namioka mount (Figure 145). A concave grating is used and the angle between the incident and diffracted beams is kept constant (approximately 70°).

5.4.5 Double Monochromators

Stray light can be reduced and higher resolution obtained by using double monochromators. An echelle spectrometer equipped with a predisperser has been used. As the two instruments are operated in tandem, the exit slit of the predisperser is the entrance slit of the monochromator.

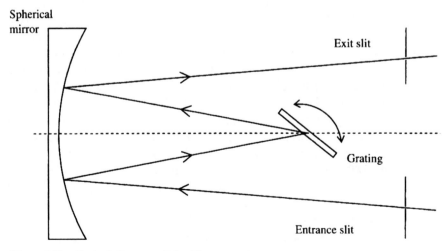

Figure 143 *Optical diagram of the Ebert mount*

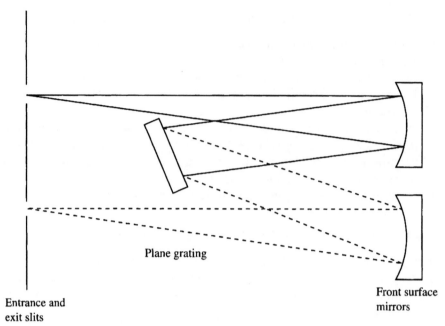

Plane grating

Entrance and
exit slits

Front surface
mirrors

Figure 144   *Optical diagram of the Czerny-Turner mount*

5.4.6   ICP Atomic Emission Fourier Transform Spectrometers

The Fourier transform spectrometer is based on the interferometer introduced in 1881 by Michelson. The basic configuration of the Michelson interferometer is shown in Figure 146. Its optical system consists of two plane mirrors at a right angle to each other and a beamsplitter at 45° to the mirrors. Radiation from the source falls on a beamsplitter, where it is divided into two equal beams for the two mirrors. A compensator (an equal thickness of support material) is placed in one arm of the interferometer to equalize the optical path length in both arms. The reflected beams from the two mirrors are recombined at the beamsplitter and emerge as a single beam. When the mirrors are positioned so that the optical path lengths of the two beams (reflected and transmitted) are equal, they will be in phase when they return to the beamsplitter and will constructively interfere. When the movable mirror is displaced by one-quarter wavelength, it will bring the two beams 180° out of phase and they will destructively interfere. As the movement of the mirror is continued in either direction, it will cause the field to oscillate from light to dark for each quarter-wavelength movement of the mirror (corresponding $\lambda/2$ changes).

When monochromatic light of wavelength λ, falls on the splitter, and the mirror is moved with a velocity v, the frequency of the signal from

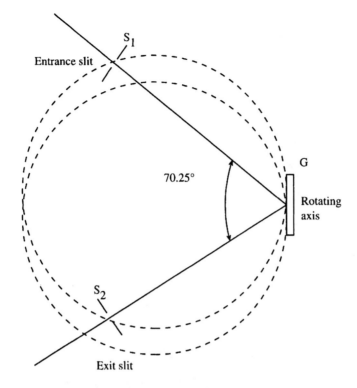

Figure 145 *Optical diagram of the Seya-Namioka mount*

the detector is $2v/\lambda$. The interferogram is obtained by plotting the signal from the detector as a function of the mirror distance. In the case of monochromatic light the interferogram is a pure cosine wave. With polychromatic light, the output signal is the sum of all the cosine waves (Figure 147). Each frequency is given an intensity modulation, which is proportional to the frequency of the incident radiation and to the speed of the moving mirror. The amplitude of the detector signal is proportional to the intensity of the incident monochromatic radiation. If the incident radiation is polychromatic, each frequency component will be transformed so that a detector output wave of unique frequency is produced for each component.

The interferogram is a Fourier-cosine transform of the spectrum. The desired spectrum is obtained from the interferogram by complex calculations involving an inverse Fourier transformation of the interferogram.

An ICP FT spectrometer has three optical inputs incorporated into a single optical axis: a He-Ne laser, a white light source (tungsten lamp),

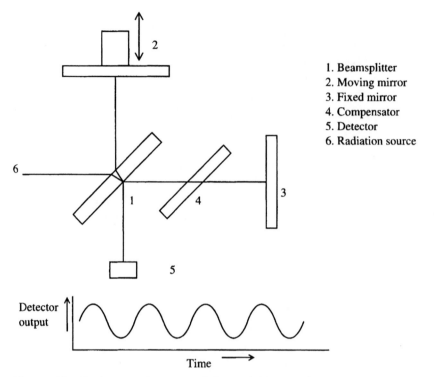

1. Beamsplitter
2. Moving mirror
3. Fixed mirror
4. Compensator
5. Detector
6. Radiation source

Figure 146   *The basic configuration of the Michelson interferometer*

and an ICP source. The laser controls the digitization sequence and measures the exact location of the moving mirror.

The main advantages of the Fourier transform spectrometer over conventional dispersive spectrometers are: (i) higher energy throughput, because no slits are required, (ii) higher optical resolution, and (iii) ability for simultaneous monitoring of all spectral information for an extended period.

5.5   INTERFERENCE EFFECTS AND BACKGROUND CORRECTION

Interference effects in plasma atomic emission spectrometry comprise: (1) nebulization interferences, (2) chemical interferences, (3) ionization interferences, and (4) spectral interferences. The degree of interference varies from one different plasma source and spectrometer used to another. However, the most significant impediment to the effective use of any plasma-AES equipment is spectral interference.

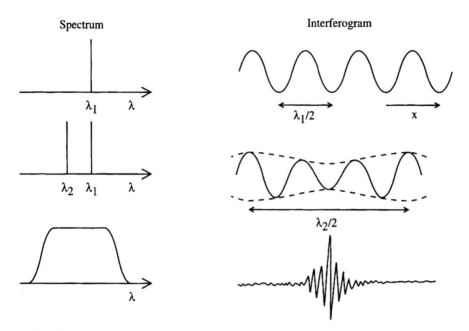

Spectrum   Interferogram

Figure 147 *Interferograms (= signal of the detector as a function of the distance of the moving mirror) of one spectral line, two spectral lines, and a continuum spectrum*

5.5.1 Nebulization Interferences

Nebulization interferences are observed if the quantity of the sample nebulized varies considerably as a function of time. Uneven nebulization may be caused by matrix salts or organic compounds and solvents by alterations in the viscosity, in the surface tension, or in the solution density (equation 46). This may also occur to some extent for solutions with high mineral acid concentrations.

Apart from this, memory effects may exist if long tubes and large vessel surfaces are used in the nebulization and desolvation system. This type of interference is not the fault of the plasma source, and can be easily avoided and controlled. For example, determination of boron may cause difficulties when an ordinary glass made spray chamber is used. This is due to attachment of this element onto surfaces of the spray chamber. To avoid this effect a spray chamber with a low dead volume should be used. In addition, a complexing agent (such as mannitol) should be added to solutions to minimize the adsorption of boron.

Interferences caused by desolvation are related to differences in volatility between the matrix and analyte, and are noticed as a change of the analyte concentration in the aerosol mist during desolvation.

The interference by copper on the determination of iodide is explained by the following reaction taking place in the nebulization chamber:

$$2Cu^{2+} + 4I^- \rightarrow 2CuI + I_2$$

Thus, the concentration of free iodine in the gas phase increases, while that in the drain solution (that leaves the aerosol) decreases. The same effect has been noted with iron(III) and other oxidizing agents.

Difficulties may occur in the analysis of organic samples, such as oils or organic solvents, and samples with high salt concentrations, if the injector tube orifice clogs with carbon or salt deposit after prolonged spraying.

5.5.2 Chemical Interferences

Due to the very high gas temperature, long residence times, and inert atmosphere in the plasma, chemical interferences caused by the formation of thermally stable compounds or radicals are unlikely, especially in ICPs. For example, the interferences caused by phosphate and aluminium in the determination of calcium appear in combustion flames, DCPs, CMPs, and MIPs, but not in conventional ICPs. However, with low power ICPs these effects might exist to some extent. In general, with increased plasma power and lower carrier gas flow rates (*i.e.* robust conditions), chemical interferences and matrix effects become smaller.

5.5.3 Ionization Interferences

Easily ionizable elements, such as alkali and alkaline earth elements, may alter the intensities of the emission lines of the analyte. This is a serious problem encountered in DCPs, MIPs, and CMPs, but causes little or no change in the intensity of the spectral lines in ICPs due to high electron densities of these sources. Figure 148 shows changes in the intensity of the iron signal line at 259 nm caused by some elements in DCP.

5.5.4 Spectral Interferences

Spectral interferences are observed in every emission source. These interferences are most important in ICPs because emission lines that might be expected to be weak or nonexistent in other sources such as flames, arcs, or sparks, are quite intense. All spectral interferences

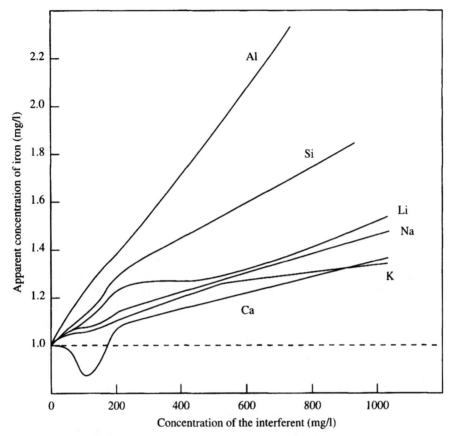

Figure 148 *Influence of some elements on the Fe emission signal at 259.9 nm in DCP-AES*
(Adapted from: L. H. J. Lajunen and R. Anttila, Finn. Chem. Lett., 1984, 63)

originate from the inherent argon spectrum, or from line and continuum spectra of atomic and molecular species entrained or injected into the plasma. Spectral interferences may be classified into four principal groups: (1) Spectral line coincidence, (2) Overlap with nearby broadened line wing, (3) Spectral continuum, and (4) Spectrometer stray light.

5.5.4.1 Spectral Line Coincidence. Direct line coincidence occurs when the monochromator of the spectrometer is not capable of separating the analyte line from the matrix line. For example, only commercial spectrometers equipped with echelle monochromators (or polychromators) are capable of routinely resolving the cadmium arsenic

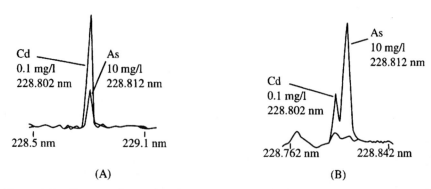

Figure 149 *Emission lines of cadmium* ($\lambda = 228.802$ nm) and arsenic ($\lambda = 228.812$) *obtained by a plasma spectrometer equipped with a conventional (A) or echelle (B) monochromator*

line pair at 228.8 nm (Figure 149). Another similar example is separation of the zinc line at 213.856 nm from the nickel line at 213.858 nm.

Little can be done to avoid spectral overlap interferences occurring at one of the fixed or preset wavelengths. Often the only solution to this problem is the separation of the analyte from the interfering matrix component by a suitable method, such as liquid-liquid extraction or ion exchange. However, with instruments offering unlimited selection of analyte wavelengths, these problems may be by-passed by moving to another wavelength that does not exhibit the interference.

5.5.4.2 Overlap with Nearby Broadened Line Wing. A strong, broadened line of a matrix element in the vicinity of the analyte line may cause spectral interference by overlapping the analyte line (Figure 150). This interference may be avoided by moving to another interference-free analyte line, by chemical separation, or by using background correction.

5.5.4.3 Spectral Continuum and Stray Light. A characteristic feature of a plasma source is a spectral continuum observed. This is due to radiative recombination of electrons, mainly with argon ions, and to lesser with other matrix elements (e.g. magnesium ions). The increase of forward power in ICP will increase the electron number density and thus increase the background emission. The background emission due to argon plasma itself is mostly compensated through the use of a solvent blank. However, spectral interference caused by one of the matrix components emitting a continuum spectrum at the analyte line wavelength may be overcome by changing the analyte line, using background correction, or by using chemical separation. For example, in the emission spectrum of aluminium an unstructured continuum begins at 212 nm and continues down below 190 nm (Figure 151). This

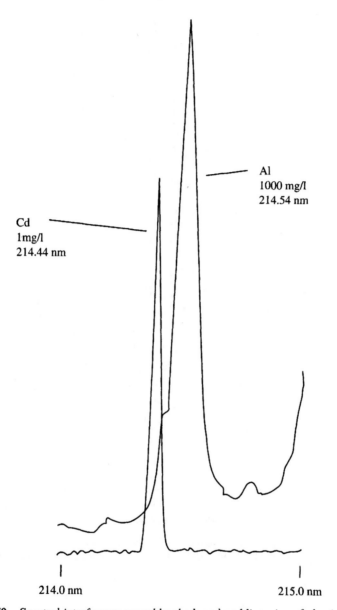

Al
1000 mg/l
214.54 nm

Cd
1mg/l
214.44 nm

214.0 nm   215.0 nm

Figure 150   *Spectral interference caused by the broadened line wing of aluminium in the*
determination of cadmium
(Adapted from: R. D. Ediger and F. J. Fernandez, At. Spectrosc., 1980,
1, 1)

causes a problem with the emission lines of cadmium (at 214.4 nm),
boron (at 208.9 nm), and tungsten (at 207.9 nm) which lie within this
region (Figure 152).

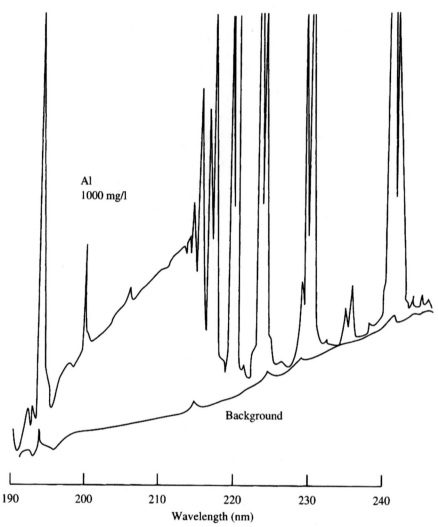

Figure 151 *Emission spectrum of aluminium between 190 and 240 nm*
(Adapted from. R. D. Ediger and F. J. Fernandez, At. Spectrosc., 1980,
1, 1)

High concentration matrix elements (e.g. magnesium, calcium or
iron) will emit intensive radiation. Due to imperfections of the optical
system, this radiation may reach the detector. The stray light raises the
background and increases the noise in the measurements. Stray light
effects may be minimized by the use of high quality optical components
and, for example, by using solar-blind photomultiplier tubes that do not
give response at higher wavelengths (above 350 nm).

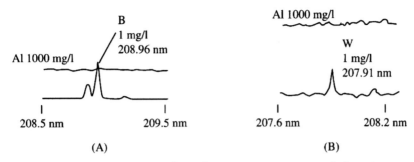

Figure 152 *Spectral interference from the continuum emission of aluminium on the determination of boron (A) and tungsten (B)*
(Adapted from: R. D. Ediger and F. J. Fernandez, At. Spectrosc., 1980, 1, 1)

5.5.5 Background Correction

In some cases the interference caused by a background spectrum line which directly coincides with the desired analyte can be avoided by background correction. For example, the OH molecular line overlaps the Al I line at 308.2 nm. Because the OH signal is relatively constant, it can be subtracted out with the blank signal. However, this will add background noise that can interfere with the determination at low analyte concentrations. This type of problem is much greater when the spectral line originates from a concomitant line which directly overlaps the analyte line, like the coincidence of the Cu I line at 213.859 mm with the Zn I line at 213.856 nm, or that of the As I line at 228.812 mm with the Cd I line at 228.802 mm. In such a case, an alternative line should be used. However, this may also be problematic if the analyte has few sensitive lines, as in the case of cadmium. When a higher resolution spectrometer, such as an echelle spectrometer, is used the lines are farther apart than their physical widths. However interference is still possible through broadened line wing overlap or stray light.

The selection of a background correction technique to eliminate or to minimize spectral interference depends on the shape of the background emission. Spectral distribution of background emission can be divided into four general cases: (i) flat background, (ii) sloped, linear background, (iii) curved background, and (iv) structured background (Figure 153).

The background correction of a simple flat background (Figure 153A) involves measuring the background intensity at some point away from the analyte peak and subtracting that value from the total signal (analyte + background). Thus, a single off-peak point measurement is sufficient.

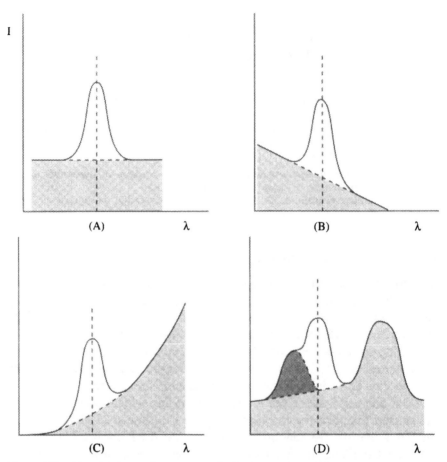

Figure 153   *Main background types: (A) flat background, (B) sloped linear background, (C) curved background, and (D) structured background (B)*

In the case of sloped background which changes in an approximately linear manner, background measurements at equal intervals off the analyte peak on either side of the peak are required (Figure 153B). Then the average of the signals, or weighted average in the case of unequal distances of the off-peak points, is subtracted from the total emission signal. Alternatively, the wavelength window may be scanned to construct a baseline beneath the analyte peak. The analyte signal is measured as the distance from the peak to the constructed baseline.

A curved background may be the result of a strong concomitant line wing near the analyte peak (Figure 153C). This is a very difficult case to correct and for example, two off-peak measurements give only an approximate correction. By multi off-peak measurements a more accurate artificial background may be generated. In many plasma instruments it is possible to scan the blank (the sample matrix without the analyte) and store the resulting spectrum. Then the spectrum of the sample is subtracted point by point from this background spectrum.

A moderately structured background with low-level maxima and minima is common for plasma AES (Figure 153D). In this case a dual-wavelength technique may be used. Measurements are made at two wavelengths where the background signals are identical, but where the analyte signals are different. Then the difference in signals at these two wavelengths is the measure for the analyte alone.

Modern plasma AES instruments equipped with echelle polychromators (Figure 142) collect large amount of spectral information simultaneously. Hence analytical results obtained for a particular element with different emission lines can be compared to detect spectral and other interferences. Sophisticated chemometric methods can be employed to correct spectral inter-element interferences observed. However, one should remind that even a simple background subtraction based on two measurements will increase the uncertainty of the net signal. If an equal variance is associated with both measurements, the standard deviation of net signal will increase by a factor of $2^{1/2}$.

5.6 ANALYTICAL FEATURES OF AN ARGON ICP

Inductively coupled plasma is by far the most common plasma source used in atomic emission spectrometry. The classic design of the spectrometers includes the radial viewing of the plasma at right angle to the plasma torch (side-on viewing). Nowadays an axial measurement arrangement, where the plasma is end-on viewed, is coming increasingly popular. With the radial viewing of plasma the optics is easily protected from overheating and salt deposition on the entrance slit and optics cause no problems. With axially viewed plasmas this may cause problems and must be taken into account when the interface between the plasma and optics is designed. The cool plasma tail must be removed from the optical axis or otherwise reduced dynamic linear range results due to absorbing atoms in the optical path. In addition, the presence of easily ionized elements may cause ionization interferences. The cool plasma tail can be deflected by a gas flow (e.g. Ar) directed at right angle to the plasma causing deformation of the plasma tail. The other possibility is to deflect the cool plasma tail by an opposite gas flow (end-on configuration). In this arrangement the cool tail is deflected away symmetrically by a shear gas that is flowing through a water-cooled interface.

Lower detection limits are usually obtained with axially viewed plasmas. However, the results obtained depend very much on the element and the emission line used. The improvement varies from two-fold improvement to approximately thirty-fold improvement when ultrasonic sample introduction is used. Low detection limits are needed, for example, when low toxic element concentrations are measured from environmental samples.

Robust measurement conditions are of prime importance when axially viewed plasma is used. Therefore, the internal diameter of injector tube employed is often larger than 2 mm. Together with a high plasma power this ensures long residence times of sample atoms and good atomization and excitation efficiency, seen as a high Mg II/Mg I ratio (> 8). Robust measurement conditions ensure also minimum interferences due to presence of high concentration sample constituents, like easily ionized elements and calcium. With robust conditions quite similar result for "difficult" samples should be obtained with both measurement configurations. For most difficult samples, for example those containing high concentrations of dissolved solids, best results are obtained with radial plasma, where the optimization of viewing height is possible. Wide linear dynamic ranges are obtained with both configurations.

5.6.1 Emission Lines Employed in ICP-AES

Inductively coupled plasma is a very efficient excitation source and many atomic and ionic lines are available for measurements. Many emission lines used are ionic lines (Table 34), but also atomic lines are employed. In order to simplify the categorization of different emission lines terms "*soft*" and "*hard*" lines are exploited. Soft emission lines are atomic lines of the elements having a first ionization potential below 8 eV. All ionic lines and atomic lines of elements having ionization potential above 8 eV are considered to be hard emission lines (Table 34). There are differences in the behavior of the two types of lines. Soft emission lines exhibit their maximum intensity < 10 mm above the load coil. Hard emission lines exhibit their maximum intensity at higher viewing heights, approximately at the same position. When the nebulizer gas flow rate is increased, the intensities of both types of lines are lowered and the maximum intensities are observed at higher viewing heights.

The changes in instrumental parameters have a very pronounced effects, when very hard or soft lines are measured (e.g. Zn II

Table 34 *Excitation potentials (EP) and ionization potentials (IP) for some emission lines used in ICP-AES*

Hard lines	λ/nm	EP/eV	(EP + IP)/eV	Soft lines	λ/nm	EP/eV
Ti II	334.941	3.70	10.53	Cu I	324.754	3.82
V II	310.230	4.00	10.74	Al I	308.215	4.02
Mg II	280.270	4.42	12.07	Mg I	285.213.	4.35
Fe II	259.940	4.77	12.67	Hg I	253.652	4.89
Mn II	257.610	4.81	12.25	Si I	251.611	4.93
Cr II	206.149	6.01	12.78	B I	249.678	4.96
Ni II	231.604	5.35	12.99	P I	213.618	5.43
Pb II	220.353	5.62	13.04	As I	193.695	6.40
Mo II	202.030	6.13	13.23			
Co II	228.616	5.42	13.30			
W II	207.911	5.96	13.83			
Cd II	214.438	5.78	14.77			
Zn II	206.200	6.01	15.40			

206.200 nm, or K I 766.490 nm). The increase in forward power considerably increases the net line intensity of Zn II line, whereas, to a lesser extent, the opposite is true for the potassium line. The stability of power source is extremely important to avoid drift, especially in the measurement of hard emission lines. The atomization and excitation efficiency of plasma can be monitored in various ways, for example by calculating the intensity ratio of magnesium ionic and atomic emission lines at 280.270 and 285.213 nm, respectively. The atomization and ionization efficiency of the plasma is considered to be good if Mg II/Mg I ratio is >8. Generally, higher ratios are obtained by increasing the forward power and reducing the carrier gas flow rate to increase the residence time of the sample constituents in the plasma. Otherwise, lower power levels and higher carrier gas flow rates increase the measurement sensitivity of soft atomic lines.

5.6.2 Optimization of Instrumental Parameters

The optimization of instrumental parameters of an ICP-AES measurement is an elaborate task that depends on the goal of optimization and the elements to be determined. Since excitation properties of various lines may vary considerably, the optimum instrumental parameters should, in principle, be selected individually for each line. The most important parameters to be optimized are the forward power of plasma, carrier gas flow rate and the viewing height of the radial plasma. Since there are likely to be interaction effects between the instrumental parameters to be optimized, one should use a proper optimization procedure (e.g. simplex optimization) instead of one-variable at a time approach.

For example, to obtain low detection limits, net signal to background ratio (SBR) should be optimized. Background emission of the argon plasma increases when the forward power is increased. Although higher net signal intensities are obtained, measurement of weak emission signals is deteriorated due to increased background noise. However, with higher power settings the small variation of plasma power affects less the net signal intensity and better precision is obtained when higher concentrations are measured.

Fortunately, compromise conditions provided by the instrument manufacturer can be employed in most analytical cases; *i.e.* sample uptake rate, carrier gas flow rate and forward power are fixed, and the viewing height of the radial plasma is optimized using an emission line suitable for a similar group of emission lines.

CHAPTER 6

Inductively Coupled Plasma Mass Spectrometry

ICP-MS can be used both for qualitative and quantitative trace elemental analysis. Isotope ratio determinations are also possible; most reliable results are obtained with multiple-collector ICP-MS instruments. In the conditions prevailing under normal analysis many elements are almost completely ionized (uncharged) in the ICP source (Figure 3). Further advantages of ICP-MS are extremely low detection limits, wide linear range (typically about 7–8 orders of magnitude), and uncomplicated spectra (Figure 154). Normally, the sample is introduced in the form of a solution into the plasma, but direct analysis of gaseous or solid samples is also possible.

6.1 INSTRUMENTATION

A comparison of the ICP-MS and ICP-AES techniques is presented in Table 35.

Most commercial ICP-MS instruments are equipped with quadrupole mass analyzers, but time-of-flight mass spectrometers or high resolution double-focusing instruments, including magnetic and electrostatic sector analyzers, are also in market. The principle of a quadrupole ICP-MS system is shown in Figure 155. The plasma torch is in a horizontal position, and it works under normal pressure. An interface cone (sampling cone) is placed between the plasma source and mass spectrometer. Ions produced in the ICP are transferred into the mass spectrometer through a small hole (about 1 mm in diameter) in the cone. At the first stage the pressure is reduced by a mechanical pump. Plasma gas flow passes through the skimmer cone to the area, where the pressure is maintained below 1 mtorr with a turbo pump(s).

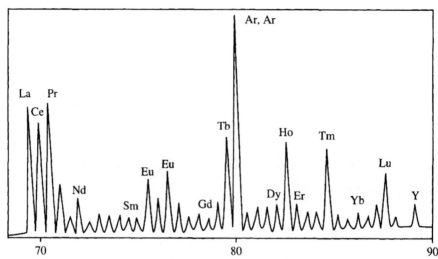

Figure 154   *Mass spectrum of univalent rare earth ions. The concentration of each rare earth element is 15 mg l^{-1}*
(Adapted from: A. L. Gray and A. R. Date, Analyst (London), 1983, **108**, 1033)

The ion beam contains mostly Ar^+ ions and trace amounts of analyte and other ions. Since the velocity of the particles in the ion beam is the same, the kinetic energy of heavier ions is much higher. Under reduced pressure electrons diffuse rapidly away from the ion beam, resulting in charge separation. Positive ions rebel each other and hence only fractions of percent of the ions are transported to the quadrupole mass filter; lighter ions are repelled more strongly since they have lower kinetic energies (space charge effect). Sensitivity of the lighter ions tends to be poorer also due to the scattering caused by collisions with neutral gas. The voltages of ion lenses have to be carefully optimized to obtain a maximum transport of various ions with different kinetic energies in to the mass analyzer. In addition, the neutral species and photons must not enter the detector to keep the background signal low. This is achieved by different means, like putting photon stops on the axis of the ion beam and focusing the ion beam so that it passes the photon stop. Quadrupole mass filter can also be positioned off-axis compared to the beam.

Quadrupole mass filter consists of four precisely machined and positioned ceramic or metal (e.g. Mo) rods. Radio frequency AC voltage is applied to opposite pair of rods connected electrically; there is a 180° phase shift between the two pairs of rods. In addition, a DC voltage is applied to the opposite pair of rods. With particular AC and DC voltages, only ions with narrow mass to charge ratio pass through

Table 35 *Comparison of the ICP-MS and ICP-AES techniques*

ICP-MS	ICP-AES
ICP	
Ionization source	Excitation source
Diagnostic tool for signal source	
CeO^+/Ce^+ intensity ratio	MgII/MgI ratio
Qualitative analysis	
m/z values	Wavelengths
Quantitative analysis	
Peak intensity (m/z)	Peak intensity (λ)
Simultaneous multi-element analysis possible	Simultaneous multi-element analysis possible
Comparison to standards, internal standardization	Comparison to standards, internal standardization
Isotope dilution method	
Isotope ratio analysis	
Possible	Impossible
Detection limits	
sub p.p.b. level	sub p.p.m. level
Spectra	
Uncomplicated	Complicated
Interference effects	
Signal enhancement and suppression effects	Signal enhancement and suppression effects
Spectral interferences	Spectral interferences
Almost free of chemical interferences	Almost free of chemical interferences

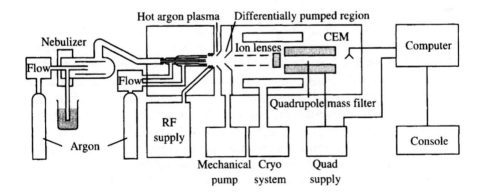

Figure 155 *Block diagram of an ICP-MS system*
(Perkin Elmer Corp.)

the quadrupole. Lighter ions are more affected by the oscillating high frequency AC voltage and heavier ions by the DC voltage. The whole mass spectrum is obtained by scanning both the voltages. The scan speed is rapid; the whole mass range can be scanned in approximately 0.1 second. Hence the speed is reasonable fast in most instances for recording transient signals obtained by sample introduction with flow injection, laser ablation, electrothermal vaporization or hyphenated chromatographic techniques.

The resolution of typical quadrupole mass filter is 1 amu (defined as peak width at 10% of its height). Improving of resolution results in reduced sensitivity, and vice versa. The mass peaks obtained are not symmetric. Low-mass abundance sensitivity is defined as a ratio of peak heights at masses $m-1$ and m. High-mass abundance sensitivity is defined similarly using peak height at masses $m+1$ and m. More tailing towards lower masses is observed with quardupole mass filters.

The detectors in ICP-MS are in principle quite similar to those used for photon detection in AAS and AES (Figure 25). Ions hitting the surface of the first dynode will release primary electrons that are accelerated to the second cathode where more electrons are released, and so on. Since the signal intensities in ICP-MS may vary at extremely large range (from less than 1 counts s^{-1} (Hz)) a pulse counting mode is used at low signal intensities. At higher concentrations analogue detection mode is used. ICP-MS instruments automatically switch the detection mode according to the signal intensity obtained.

The instrumental detection limits obtained with solution nebulization for more than 60 elements, including carbide forming elements and lanthanides, are extremely low (Table 49). For most elements detection limits are below 10 ng l^{-1} (p.p.t.). Instrumental detection limits for bromine, chlorine, phosphorus and silicon lie between 0.1 and 1 µg l^{-1} (p.p.b.). Sulphur is preferably measured by ICP-AES than by ICP-MS. However, method detection limits obtained with ICP-MS are often considerably higher than instrumental detection limits, for instance due to polyatomic interferences encountered in the analysis of real samples (Tables 37–39). The situation is especially troublesome if the analyte element has only one isotope available for determination.

Avoidance of contamination in ICP-MS analysis is extremely important since very low analyte levels can be measured, or otherwise the accuracy of the analysis results at ultra trace levels is deteriorated.

6.2 SAMPLE INTRODUCTION

ICP-MS use the same sample introduction techniques that have been used for many years in optical ICP-AES. When ordinary solution nebulisation is used for sample introduction, an excess solvent load is a problem in ICP-AES and ICP-MS that affects the analytical properties of the ICP source. In ICP-MS, the formation of solvent based polyatomic ions causes problems due to overlap with analyte isotopes. Since water is the most common solvent used, many interferences due to oxygen based ions exist (Table 39). Therefore various desolvation systems (e.g. a cooled spray chamber) are frequently employed in ICP-MS.

Normally, solid samples are dissolved by a suitable method (wet or dry digestion, fusion, pressure dissolution). However, dissolution procedures may lead to contamination and increase the dissolved solid content of the final sample solution. Highly dissolved solids may cause interface cone blockage and formation of polyatomic ions from solvent matrix and plasma gases.

Slurry nebulization has been used as an alternative technique for solid sample introduction in ICP-MS where sample dissolution is not needed. It has been successfully applied for analysis of geological and soil samples, and industrial catalysts.

To avoid above mentioned problems due to solvent and high dissolved solids, several dry sample introduction techniques have been applied to analyse solids directly without dissolution.

6.2.1 Laser Ablation

Laser ablation (LA) minimizes problems existing in solution introduction. Using laser ablation it is possible to analyse samples in a dry plasma with minimum oxygen- and hydrogen-related interferences and better plasma operation characteristics. Background and detection limits are low, and sensitivity high. In the laser ablation technique a portion of a solid sample is volatilized by the laser, and transported usually by an argon flow to the ICP torch. The volatilized material is atomized and ionized in the ICP for mass spectrometric elemental analysis.

A laser ablation sample introduction system includes a laser (operating usually in Q-switched mode), a deflection or focusing optics unit, an ablation chamber (containing the sample and generated aerosol), and a connection between the ablation unit and the ICP to transport the aerosol into the plasma (Figure 156).

A Nd:YAG laser (Neodymium doped Yttrium Aluminium Garnet rod) is most often employed. This laser has a fundamental wavelength of 1064 nm. With optical frequency quadrupling or quintupling wavelengths 266 nm and 213 nm (often used in commercial systems) are

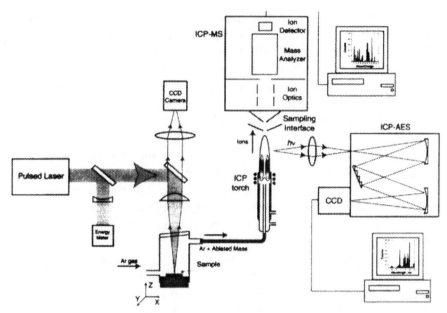

Figure 156 *Schematic of a laser ablation sample introduction system*
(Adapted from: R. E. Russo, X. Mao, H. Liu, J. Gonzalez and S. S. Mao, Talanta 2002, **57**, 425)

obtained. Generally, shorter wavelengths are preferred due to better absorption of radiation to sample when compared to visible or infrared radiation. Also smaller ablation spot sizes are obtained with UV lasers. There is also evidence that the use of shorter laser wavelengths results in smaller ablated particles which are more easily atomized and ionized in the plasma.

In laser ablation various chambers for different types of samples have been designed to analyse rod and disc-shaped samples, and powders. For example, a disc ablation chamber system accommodates a range of disc sizes including the typical metal industry standards used in spark emission spectrometry. The internal volume of ablation chamber is typically below 100 ml. Smaller internal volumes allow a rapid purging of the chamber and the dilution of ablated material is less if very small sample details (e.g. inclusions) are studied. The sample aerosol generated is transported into the plasma source through a PVC tube. In order to keep the distance between the ablation chamber and the plasma as short as possible, the chamber may be placed directly below the ICP.

Elemental fractionation is a problem in LA analysis. This means that the ablated material analyzed by ICP has a different average composition to that of the bulk sample. There are several processes that may lead to elemental fractionation. These can be traced to (i) properties of elements ablated; (ii) the influence of ablation crater geometry; (iii) the effects due to aerosol transportation and (iv) processes occurring in the plasma.

It has been found out that laser wavelength has a significant effect on the elemental fractionation. The use of long wavelength IR lasers results in significant fractionation. This is partly due to the fact that thermal processes prevail during ablation when longer wavelengths are used. A plasma is formed above the surface of ablated sample that decreases the interaction between the sample and laser beam. With UV lasers (e.g. those using wavelength of 213 nm) the laser beam is absorbed to a lesser extent by the plasma and the direct interaction (bond breaking of the sample compounds) is more efficient. In contrast, longer wavelengths tend to be absorbed more strongly by the surface plasma, resulting in the temperature increase above the sample surface. This in turn results to sample melting, and selective evaporation of the sample components, and thus to elemental fractionation.

Fryer et al. have showed that elements can be categorized according to their relative fractionation tendencies using Goldschmidth's classification (lithophile, siderophile, and chalcophile elements). Thus, for example, alkaline earth elements, Al, Sc, Y and rare earth elements are lithophile and show similar ablation behavior. This is an important observation, since for instance Ca can be used as an internal standard in the determination of these elements by LA-ICP-MS if suitable calibration standards are not available.

It has been found out that also ablation crater shape has a significant effect on the fractionation. The exact mechanism of crater influence has not been established. However, if the crater is too deep, material is not efficiently removed, resulting in fractionation. As a thumb of rule, crater depth/diameter ratio should not be greater than 3 or fractionation may occur.

Elemental fractionation may occur also during the vapour transportation to the plasma. In general, when shorter laser wavelengths are used for irradiation, this will result in smaller ablated particles that are more efficiently transported to plasma. Generally, the elemental composition between individual particles depends on the size of ablated particles. Larger particles, for instance, may settle down due to gravitational force during transportation and hence the average composition of the vapour entering plasma is different to that of bulk material. There is also evidence that larger particles (if entering the plasma) are incompletely atomized and ionized and hence elemental fractionation may occur. Also some matrix effects in the plasma are observed depending on the composition of the main matrix of ablated particles.

Several laser parameters must be carefully controlled to obtain analysis data that fit for the purpose of analysis. In addition to (fixed) wavelength used, several other parameters must be optimized. These include laser pulse energy, laser pulse energy density and line width (controlled by beam expander), and the pulse repetition rate (typically between 1–20 pulses s^{-1}). Spot widths may range form >100 μm (in bulk analysis) to few micrometers (attainable with UV lasers). For instance, with higher beam expander settings larger spots sizes are obtained, but lower crater depths result due to lower energy density of the laser beam. In addition, the analyst selects the scan speed for a raster line if larger area is studied (e.g. in bulk analysis). Scan speeds between 10 μm s^{-1}– 100 μm s^{-1}, for instance, may be selected. Hence the amount of material ablated during an analysis (and spatial information gained) depends on many experimental variables that must be carefully optimized. These variables are also related to the number of isotopes selected for analysis, *i.e.* what amount of time signal is gathered per isotope (if fast sequential ICP-MS instrument is used).

Since many variables affect the amount of sample ablated and in addition aerosol transport efficiency and ICP related aspects (background signal, isotope sensitivity, etc) must also be taken into account, detection limits and precision obtained are very dependent on particular analytical situation. However detection limits from sub p.p.m. level (high spatial resolution) to sub p.p.b. level (low spatial resolution) are usually obtained. In Table 36, examples of detection limits are given, obtained for NIST 681 platinum sample.

Calibration is always a major problem in any technique used for the direct analysis of solids. Several approaches for calibration in

Table 36 *Detection limits (3σ) obtained with solution base calibration for Platinum (NIST 681) with laser ablation ICP-MS* (Source: J. S. Becker, Spectrochim. Acta 2002, **57B**, 1805)

Element	Detection limit (μg g^{-1})
Cu	0.04
Rh	0.003
Ag	0.002
Pd	0.003
Id	0.001
Pb	0.003
RSD (%)	2–10
Accuracy (%)	<8
Analysis time (h)	~1 h

LA-ICP-MS are possible. The most straightforward calibration is carried out with certified reference materials (such as NIST glasses or metal samples) that are alike to samples to be analyzed. The ablation characteristics of the standards and samples must be similar (matrix matching). External, matrix-matched calibration standards can also be in-house prepared, although this may be difficult since the exact composition of the samples is not always known. For instance, standard compounds are added to a powdered standard matrix, the mixture is homogenized carefully and pressed to a disk. External standards can also prepared by fusion.

Internal standardization is often applied together with external calibration. The internal standard added must be homogenously distributed in the sample matrix. In addition, the ablation behavior and matrix effects occurring in the plasma must be similar between the internal standard and the analytes measured. This may be difficult to achieve when only one internal standard, such as indium, is used. Sometimes it is possible to use elements existing naturally in the sample as an internal standard. In this case the concentration of the element(s) must be accurately known, or determined by an independent method.

Calibration using standard solutions is also possible. This may include dual sample introduction system where the standard aerosol leaving the spray chamber is mixed with the ablation vapours carried by argon form the ablation chamber. During sample ablation solution blank is introduced. Desolvation system should be used or otherwise the advantage of dry plasma obtained with laser ablation is lost. Desolvatated aerosol may be also forced through the ablation cell. Matrix effects are likely to occur with solution calibration since the properties of the ablated particles and desolvatated standard particles might be considerably different. Elemental fractionation is neither taken into account, making solution calibration procedures less reliable. Although calibration is a major challenge in LA-ICP-MS, accuracies better than 10% can be obtained with external calibration when solid calibration standards are used.

Laser ablation ICP-MS can be used for analysis of a great variety of solid samples: conducting and nonconducting materials, solids, and powders. In semiconductor industry, for example, impurity levels allowed are extremely low and the determination of impurities at pg g^{-1} level is almost impossible due to contamination if samples must be dissolved before analysis. Therefore direct LA-ICP-MS is a very important analytical technique in these special cases. In addition, due to the focusing characteristics of the laser beam, investigations of very small localized areas in the sample surface are possible and LA-ICP-MS also offers spatial information from heterogeneous samples (e.g. depth profiling).

In the bulk analysis an average concentration of a solid sample is determined without sample digestion. Large area is usually scanned (large beam width or rastering) to obtain representative result. The material must be homogeneous to obtain accurate results. Some sample types (e.g. metal alloys) may fulfill this requirement. Laser sampling ICP-MS has also been applied successfully as a semi-quantitative survey tool for a rapid multi-element analysis of geological samples. The samples are pulverized to assure homogeneity. Some paraffin wax is mixed with the sample. The mixture is then placed in an aluminium cup and pressed into a wafer. The fusion of the sample to form glass is also possible.

LA-ICP-MS is a rapidly growing analytical field that has many important applications, especially when information from spatially restricted small areas is wanted. Application areas include fluid inclusions, analysis of layered systems, study of surface contamination, analysis of very pure materials, analysis of biological samples and study of archaeological specimens, etc. Macroscopically, LA-ICP-MS is almost non-destructive method in the investigation of valuable samples.

6.2.2 Arc Nebulization

Arc nebulization can be used to produce an aerosol from a conducting solid sample. The aerosol is then introduced into the plasma for elemental or isotope ratio analysis. Powdered samples can be mixed with graphite and pressed to get a conducting pellet. The only matrix related interfering ions detected are ArC^+, ArO^+ and Ar_2^+. A precision of about 1% has been reported for isotope ratio determinations.

6.2.3 Electrothermal Vaporization (ETV)

This technique uses an electrothermally heated graphite furnace, a carbon rod, or metal filament to atomize the sample. The sample is placed in the form of a solution or a solid in the atomizer.

A transport efficiency of greater than 80% has been achieved by a rhenium filament ETV system with a glass housing geometry. Optimum sensitivity is obtained with a moderate heating rate and it is dependent on the carrier gas flow.

A controlled electrical current is applied to the graphite furnace, which subsequently undergoes resistive heating. The heating cycle typically comprises three controlled stages: for example, drying at 80°C, ashing at 300–450°C, and vaporization at temperatures as great as 2800°C using argon as the carrier gas.

The requirements for ETV-ICP systems (both ICP-AES and ICP-MS) differ significantly from those for ETA-AAS. When using ETV-ICP, it is necessary to introduce a volatile sample into the gas stream. The atomization stage is then performed in the ICP unit. This is the reason why the technique is referred to as ETV in the case of ICP-AES and ICP-MS, but ETA (electrothermal atomization) in the case of AAS. The chemical matrix effects observed in AAS are negligible using ETV-ICP, whereas the transport efficiency of the sample from the ETV unit to the ICP unit is critical to the analytical performance in ETV-ICP-MS.

Using ETV-ICP-MS detection limits improved by a factor of 10 to 10^3 compared with pneumatic nebulization methods. Detection limits using pneumatic nebulizers with ICP-MS are typically 1 to 100 ng l^{-1}, and those for ETV-ICP-MS 1 to 10 pg l^{-1} (Table 49). This is due to more efficient sample introduction (the nebulization efficiency of pneumatic nebulizers is only 1 to 2%) and removal of the solvent matrix by thermal pretreatment. Thus, using this technique it is possible to analyse samples containing high levels of dissolved salts. Also the ability to operate with very small sample volumes (10 to 100 µl) is useful, especially when the amount of sample is limited.

Use of a graphite furnace is limited in practice to elements which volatilize from graphite at temperatures below 2800°C. Some elements also tend to react with carbon to form vapour-phase carbides at high temperatures. This effect can be overcome by the addition of Freon to the argon carrier gas (a flow rate of 0.1 to 0.5 ml min^{-1}). The elements react with the liberated fluorine radicals to form volatile fluoride compounds instead of carbides. Sensitivity is markedly increased by improved transport efficiency and reduced memory effects.

In addition to the single-element determinations at very low concentration levels, a major advantage of ICP-MS is in multi-element analysis.

6.3 INTERFERENCE EFFECTS

Solid deposition on the interface cone and skimmer is a problem in ICP-MS. Deposition decreases with increased power of the ICP torch. Major interference effects in ICP-MS may be divided into two groups: (6.3.1) signal enhancement and suppression effects, and (6.3.2) spectral interferences.

6.3.1 Signal Enhancement and Suppression

Instrumental parameters, especially forward power, nebulizer gas flow rate and sampling depth have a significant effect on the responses of various elements in ICP-MS. When liquid sample is introduced into the plasma, desolvation, vaporization, atomization and ionization must happen in a very short time period. Nebulizer gas flow rate has a marked effect on these processes. If the flow rate is increased, maximum ion concentration is observed more faraway from the coil. If the plasma power is increased and the nebulizer flow rate is kept constant, the maximum intensity is observed when ions are sampled more closely to the load coil. Due to variations in ionization energies, stability of oxides, etc., the instrument should be optimized for a particular element in the most demanding applications. The sampled ions have approximately the same velocity as the plasma gas when they enter to vacuum area. Hence the kinetic energies of various ions differ due to their different masses. The optimum voltages of ion lenses depend on the mass of the ions measured. Autotune option is currently available in instrument operation software to automatically tune the instrument in routine work.

Various diagnostic tools are available for instrument user to check the functioning of the instrument and for comparison to manufactures recommendations. These include CeO^+/Ce^+ ratio and Ba^{2+}/Ba^+ ratio. Most manufacturers prefer to use 27 MHz plasma instead of 46 MHz, since hotter plasma is obtained with the lower frequency.

Like in ICP-AES technique, the sample introduction is a critical step that affects sensitivity and matrix effects observed. In principle, the sensitivity is increased, if more sample aerosol is introduced into the plasma in a time unit (e.g. ultrasonic nebulization). However, ionization of the elements is suppressed due to cool down of plasma, until a desolvation system is used. ICP-MS instrument is more sensitive to sample constituents than ICP-AES. Sample introduction and plasma itself is quite alike in both of these techniques. ICP-MS is more prone to dissolved solids in samples. The gradual blocking of orifices in sample and skimmer cones can be avoided/diminished by diluting the samples (if possible) or by using flow injection technique in sample introduction. Oxygen must be added to carrier gas if organic solvents are analysed to prevent of carbon deposit build up in the cones.

Overall instrument response as a function of mass may be changed by certain concomitant elements. This same phenomenon also exists in ICP-AES, but in the case of ICP-MS it is often a more serious problem. For instance, the 1000 mg l^{-1} sodium concentration in the matrix will cause a 20% suppression of cobalt and bismuth signals when the concentration of these elements is 10 mg l^{-1}. On the other hand, large amounts of aluminium and phosphate do not affect the intensity of the calcium line. The degree of signal enhancement or suppression is directly dependent on the nature and concentration of the matrix compounds.

The matrix elements have a marked effect on the response of analyte ions. Elements with low ionization energies will cause the greatest suppression of signal intensities, and the change in the signal intensities will depend more on the concentration of the interferent than the interferent/analyte concentration ratio. This is due to lower transmission of analyte ions in the presence of large amounts matrix ions (space charge effect). The effect is most serious when light elements are measured in the presence of heavy matrix ions (Figure 157). Systematic errors due to matrix effects can be avoided by dilution of samples or by using matrix matched standards, if possible. Internal standardization is also a common practice in ICP-MS to compensate matrix effects. An internal standard has been found most effective when it is close in mass and ionization energy to the analyte. Hence several internal standard elements must be used if light and heavy elements are measured from the same sample.

6.3.2 Spectral Interferences

Spectral interferences are due to overlap of two elemental isotopes having the same nominal mass (isobaric interference), e.g. $^{63}Cu^+$ and $^{63}Ni^+$) or due to overlap of analyte isotope with polyatomic ions (containing usually 2 or 3 atoms) formed in the ionization source (e.g. $^{56}Fe^+$ and ArO^+). In practice spectral interferences are common, since most ICP-MS instruments are equipped with a quadrupole mass analyzer having a unit mass resolution. Spectral interferences can be serious if the concentration of analyte is low and the signal of an interferent is high, like in the case of argon based molecules. Common problem encountered in semiconductor industry is the determination of low levels of K, Ca and Fe by quadrupole ICP-MS. The difficulties are due to the following overlaps: $^{39}K^+$ (abundance 93.3%) and $^{38}Ar^1H^+$; $^{40}Ca^+$ (96.9%) and $^{40}Ar^+$; ^{56}Fe (91.7%) and $^{40}Ar^{16}O^+$.

The situation is even more complicated if only one isotope is available for measurement. The determination of low levels of arsenic ($^{75}As^+$, abundance 100%) is hampered by $^{75}ArCl^+$. This interference is likely to be serious, if hydrochloric acid is used in sample preparation. The interferences due to presence of high concentration of matrix element M in sample may result in formation of MO^+ and MOH^+ ions.

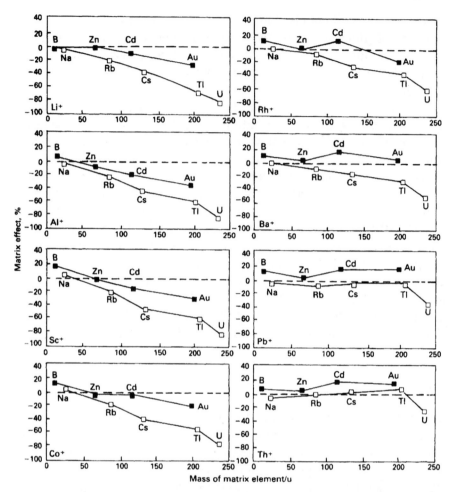

Figure 157 *Effects of some matrix ions (0.0042 mol l⁻¹) on the signals of various analyte*
elements (1 mg l⁻¹)
(Adapted from: S. H. Tan and G. Horlick, J. Anal. Atom. Spectrom.,
1987, **2**, 745)

Table 37 lists polyatomic ions formed by mineral acids commonly
used in sample preparation and element ions with the same mass/charge
ratios. Table 38 summarizes spectral interferences originating from the
plasma and carrier gases, and Table 39 lists spectral overlaps of various
MO^+ ions.

6.3.2.1 Elimination of Spectral Interferences. Most straightforward
strategy to avoid spectral interferences is to improve the resolution of
the mass analyzer; e.g. using a double focusing mass analyzer instead
of a low resolution instrument.

Table 37 *Spectral interferences caused by polyatomic ions formed by mineral acids*
(Sources: R. S. Houk, Anal. Chem., 1986, **58**, 97A; J. W. McLaren, D. Beauchemin, and S. S. Berman, ICP Info Newsl., 1987, **13**, 34)

Acid	Ion	m/z	Overlapping ion (Natural abundance,%)	
HNO₃	N⁺	14	—	
	ArN⁺	54	$^{54}Fe^+$	(5.8)
			$^{54}Cr^+$	(2.3)
HCl, HClO₄	Cl⁺	35, 37	—	
	ClO⁺	51	$^{51}V^+$	(99.7)
		53	$^{53}Cr^+$	(9.6)
	ClOH⁺	52	$^{52}Cr^+$	(83.8)
		54	$^{54}Cr^+$	(2.3)
		54	$^{54}Fe^+$	(5.8)
	ClO₂⁺	67	$^{67}Zn^+$	(4.1)
		69	$^{69}Ga^+$	(60)
	Cl₂⁺	70	$^{70}Zn^+$	(0.8)
		70	$^{70}Ge^+$	(20.5)
		74	$^{74}Ge^+$	(36.5)
		74	$^{74}Se^+$	(0.9)
	ArCl⁺	75	$^{75}As^+$	(100)
		77	$^{77}Se^+$	(7.6)
H₂SO₄	S⁺	32, 33, 34	—	
	SO⁺	48	$^{48}Ti^+$	(74.0)
		49	$^{49}Ti^+$	(5.5)
		50	$^{50}Ti^+$	(5.2)
		50	$^{50}Cr^+$	(4.3)
		50	$^{50}V^+$	(0.3)
	SO₂⁺	64	$^{64}Zn^+$	(48.9)
		64	$^{64}Ni^+$	(1.2)
		65	$^{65}Cu^+$	(30.9)
		66	$^{66}Zn^+$	(27.8)

Negative ion mode has been investigated as a means of determining elements that are difficult or impossible to determine under normal operating conditions (e.g. $^{32}S^+$ and $^{40}Ca^+$) and those which do not form positive ions in the ICP (for example, fluorine). However, very high background noise makes this technique impractical in spite of the possibility of detection limits being less than 1 mg l⁻¹. Some of the severe spectral interferences of the argon ICP, such as $^{40}Ar^+$ on Ca or $^{80}Ar_2^+$ on Se, could be overcome by the use of helium MIP.

In commercial instruments, the interferences due to isobaric overlaps are, in principle, easily corrected by mathematical equations provided by the operating software. However, one has to be aware that the isotope used for correction is free of molecular ion interferences, or otherwise overcorrection will occur.

Table 38   *Spectral interferences caused by diatomic ions originating from the
plasma and carrier gases*
(Source: R. S. Houk, Anal. Chem., 1986, **58**, 97A)

	m/z	*Overlapping ion (Natural abundance, %)*	
N_2^+	28	$^{28}Si^+$	(92.2)
NO^+	30	$^{30}Si^+$	(3.1)
O_2^+	32	$^{32}S^+$	(95.0)
	34	$^{34}S^+$	(4.2)
ArO^+	52	$^{52}Cr^+$	(83.6)
	54	$^{54}Cr^+$	(2.3)
	54	$^{54}Fe^+$	(5.8)
	56	$^{56}Fe^+$	(91.7)
	58	$^{58}Ni^+$	(67.8)
Ar_2^+	72	$^{72}Ge^+$	(27.4)
	74	$^{74}Ge^+$	(36.7)
	74	$^{74}Se^+$	(0.9)
	76	$^{76}Se^+$	(9.0)
	78	$^{78}Se^+$	(23.5)
	80	$^{80}Se^+$	(49.8)

Since the spectral interferences originate from sample and/or plasma
constituents, interferences can be reduced by affecting the chemistry of
the sample or changing the instrumental settings. Major sample component, such as chloride, causes often severe spectral interferences.
These major components are introduced through sample preparation
(e.g. HCl or $HClO_4$, Table 37) or they may exist in the original sample
matrix (chloride in sea water or biological fluids). In the analysis of
bulk materials, like metal alloys or reagent chemicals, very high matrix
element concentrations may occur. The interferences can be removed by
separating the analytes or interferents from the sample matrix. Metal
ions can be separated through complex formation and liquid-liquid or
solid phase extraction. Chloride can be removed by anion exchange, etc.

Appearance of interferences due to oxide and hydroxide molecules is
likely to cause problems since aqueous solutions are usually introduced
into the plasma: some examples are given in Table 39. To minimize the
formation of the interfering species water load to plasma should be
minimized. Thermostated and cooled spray chamber are currently
available in commercial instruments. Special desolvation systems are
also very useful. Introduction of water vapour can be eliminated by the
use of alternative sample introduction techniques described earlier. For
instance with laser ablation, electrothermal vaporization and hydride
generation practically dry sample aerosol is obtained. Matrix modification can be employed with electrothermal vaporization to further
modify the properties of the aerosol.

Table 39 *Spectral interferences caused by* MO^+ *and* MOH^+ *ions*
(Sources: J. W. McLaren, D. Beauchemin, and S. S. Berman, Anal.
Chem., 1987, **59**, 610; C. W. Mcleod, A. R. Date, and Y. Y. Cheung,
Spectrochim. Acta, 1986, **41B**, 169)

Analyte ion (Natural abundance, %)		m/z	Interfering ion
Ni$^+$	(68.3)	58	$^{42}Ca^{16}O^+$
	(26.1)	60	$^{44}Ca^{16}O^+$
	(1.1)	61	$^{44}Ca^{16}OH^+$
	(3.6)	62	$^{46}Ca^{16}O^+$
Cu$^+$	(69.2)	63	$^{46}Ca^{16}OH^+$
Co$^+$	(100)	59	$^{43}Ca^{16}O^+$
Zn$^+$	(48.6)	64	$^{48}Ca^{16}O^+$
Cd$^+$	(12.8)	111	$^{94}Zr^{16}OH^+$
	(12.8)	111	$^{95}Mo^{16}O^+$
Sb$^+$	(42.7)	123	$^{91}Zr^{16}O_2^+$

The occurrence of polyatomic ions is related to operating parameters and instrument design. The amount of interfering polyatomic ions can be reduced by optimization of plasma parameters (nebulizer gas flow rate, forward power, sampling orifice dimensions and cleanliness of interface surfaces). Sometimes the use of very low forward power and high nebulizer gas flow rate is advantageous ("nonrobust" conditions in ICP-AES). "Cold plasma" is formed when a power $\leq$ 600 W and nebulizer gas glow rate $\geq$ 1.0 ml min^{-1} are used (a grounded electrical shield is often placed between the load coil and the plasma torch). With cold plasma the molecular interferences due to Ar^+, ArH^+ and ArO^+ are reduced with several orders of magnitudes, allowing the interference free determination of K, Ca and Fe at very low concentrations. However, cold plasma conditions suit only for elements having low ionization potentials. The sensitivity of elements having ionization potentials above 6 eV is reduced. In addition, chemical interferences due to oxide formation are very likely for the elements having high M-O bond energies. Cold plasma conditions are also susceptible to matrix effects. At normal operating conditions the electron density of the ICP source is very high and the ionization of matrix elements has a negligible effect on the ionization equilibrium of analyte. However, when the plasma temperature is very low the easily ionized matrix elements cause suppression of analyte element ionization, resulting in reduced sensitivity. Therefore, careful matrix matching of standards is essential or, preferably, "clean" samples should be analyzed.

The latest technical improvement in ICP-MS to overcome spectral interferences is the introduction of collision/reaction cells. Collision/

a)

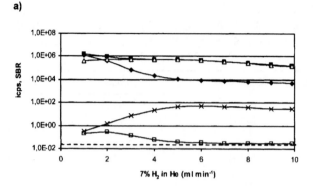

b)

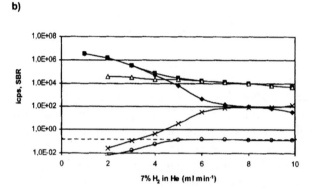

c)

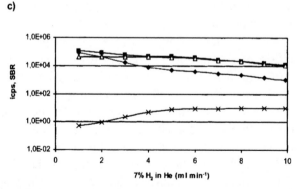

Figure 158 *Total analyte and background signals, signal-to-background ratio (SBR), net signal and isotope ratios at a) m/z 56, b) m/z 80 and c) m/z 75 as a function of the hydrogen collision gas flow rate (◆ =background, ■ =total signal, ✕ = SBR, △ =net signal, □ = $^{57}Fe/^{56}Fe$ ratio, ○ = $^{77}Se/^{80}Se$ ratio, —— = expected isotope ratio). Signals were measured using 10 μg l^{-1} standard solutions and blank solutions (2% v/v HNO$_3$ for Fe and Se; 5% v/v HCl for As)*

(Adapted from: M. Niemelä, P. Perämäki and H. Kola, Anal. Chim. Acta 2003, **493**, 3) reaction cell is placed before the quadrupole mass filter and is pressurized with a reacting gas (if necessary, the cell can be evacuated and the instrument is functioning like an ordinary ICP-MS instrument). Various reacting gases are used, hydrogen and ammonia being common examples. Occasionally, a buffer gas having low reactivity (e.g. helium) is also added to the cell. Due to collisions with the cell gases the analyte ions loose they energy and collisional focusing effect may occur, resulting in sensitivity increase. However, due to scattering, the sensitivity of lighter elements is generally somewhat deteriorated.

Although there are in principle many possible pathways for interference elimination, chemical reactions are thought to be prevailing processes. Some examples are given below:

$$Ar^+ + H_2 \rightarrow H_2^+ + Ar \qquad \text{(Charge transfer)}$$
$$Ar^+ + H_2 \rightarrow H + ArH^+ \qquad \text{(Hydrogen atom transfer)}$$
$$ArH^+ + H_2 \rightarrow H_3^+ + Ar \qquad \text{(Proton transfer)}$$

Figure 158 shows some results obtained with H_2 as a collision cell gas in the determination of ^{56}Fe, ^{80}Se and ^{75}As.

With the use of collision/reaction cells optimal plasma conditions can be used to obtain a robust plasma and good sensitivity. Inside the collision/reaction cell there is a multipole (quadrupole, hexapole or octapole), that is used for focusing the ion beam. Although the primary reaction is very efficient in eliminating the interfering molecule, the reaction products are prone to further reactions, for example with the cell gas containing trace amounts of impurities. Therefore there is a high risk that many new interfering molecules appear. Different approached are used to get rid of these potential new interferents. In a guadrupole based cell the multipole acts as mass filter and the unwanted side products are rejected. In hexapole and octapole the stability region of multipole is diffuse and mass filtering is not possible. In these devices kinetic energy discrimination is used. The new species formed in the cell have a lower kinetic energy than the ions coming from the plasma. By careful adjustment of cell bias the interfering molecules are discriminated and only analyte ions transmitted to the quardupole.

CHAPTER 7

Atomic Fluorescence Spectrometry

Atomic fluorescence spectrometry (AFS) is based on the excitation of gaseous atoms by optical radiation of suitable wavelength (frequency) and the measurement of the resultant fluorescence radiation. Atomic fluorescence is, thus, in principle the opposite process to atomic absorption. Each atom has a characteristic fluorescence spectrum. The wavelength of the fluorescence line may be the same, greater, or smaller than the wavelength of the excitation line.

Atomic fluorescence transitions may be divided into three types: *(a)* resonance fluorescence (the excitation and fluorescence lines have the same upper and lower energy levels), *(b)* direct line fluorescence (the excitation line and fluorescence line have the same upper energy level, but different lower energy levels), and *(c)* stepwise line fluorescence (the excitation line and the fluorescence line have different upper energy levels) (Figure 159).

The resonance fluorescence is the most common form, and in this case the excitation and fluorescence lines have the same wavelength. The resonance fluorescence lines may originate from the ground state or from an excited state.

If the wavelength of the fluorescence line is greater than that of the excitation line, the effect is called Stokes direct line fluorescence. In the reverse case this effect is called anti-Stokes direct line fluorescence. All these lines may originate either from the ground state or from an excited state.

In the case of the stepwise line fluorescence, the effect is divided into Stokes and anti-Stokes stepwise fluorescence depending on the wavelength (energy) relationships. A thermally assisted process may take place, if after radiation excitation, further collisional excitation occurs.

The intensity of atomic fluorescence depends on the intensity of the incident radiation source, concentration of the analyte atoms in the

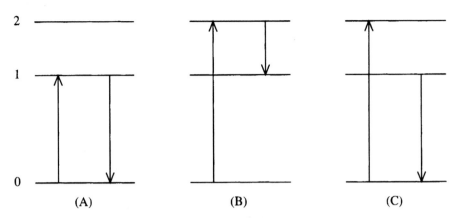

Figure 159 *Types of atomic fluorescence: (A) resonance fluorescence, (B) direct fluorescence, and (C) stepwise fluorescence (0 = ground state, 1 = first excited state, 2 = second excited state)*

ground state, absorption efficiency of the incident radiation, and degree of self-absorption in the atomization cell.

The excitation source can be a continuum or a line-like radiation source. Research on atomic fluorescence spectrometry has been connected with the examination of intense radiation sources such as electrodeless discharge lamps and lasers. Various flames, plasmas, and furnaces have been employed as atomizing devices.

The first atomic fluorescence spectrometer was constructed by Winefordner and Vickers in 1964.

7.1 THE RELATIONSHIP BETWEEN FLUORESCENCE AND CONCENTRATION

The intensity of fluorescence radiation between the ground state, j and an excited state, i is described by the following equation:

$$I_F = (l/4\pi)\, Y_{ij} I_{v_{ji}} \int k_v \, dv \tag{83}$$

where l is the distance between the excitation cell and detector, Y_{ij} is the quantum yield, $I_{v_{ji}}$ is the intensity of the excitation radiation at the frequency v_{ji}, and $\int k_v dv$ is the integrated absorption coefficient over the absorption line. The integrated absorption coefficient is proportional to the concentrations of the states j and i and to their statistical weights, the intensity of exciting photon, and the Einstein absorption coefficient, B_{ji}.

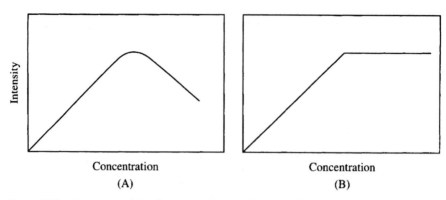

Figure 160   *Intensity of the fluorescence as a function of analyte concentration when using (A) a line-like radiation source, and (B) a continuum radiation source*

The fluorescence intensity (I_F) is in a linear relationship between the intensity of the incident radiation and the quantum yield of fluorescence as long as $I_{v_{ji}}$ is sufficiently below the level of saturation.

The fluorescence equation is dependent on the nature of the excitation source. If a line-like radiation source is employed, a term which takes into consideration the half-width of the excitation line with respect to the half-width of the absorption line is included. In the case of a continuum source, a term for the product of the Doppler half-width and the continuum incident radiation must be added to the equation.

Figure 160 represents the intensity of fluorescence as a function of the concentration of the analyte when employing either a continuum or a line-like radiation source. Using a line-like radiation source, the intensity may curve against the concentration axis with increasing analyte concentration. Thus, two concentration values may correspond to one intensity value. The decrease in intensity is caused by self-absorption of the analyte atoms in the atomization cell at high concentrations.

7.2   INSTRUMENTATION FOR ATOMIC FLUORESCENCE SPECTROMETRY

A schematic structure of a simple AFS instrument is shown in Figure 161. The main units of the instrument are: (i) excitation source, (ii) atomization cell, (iii) wavelength selection unit, and (iv) detection and readout devices. Sample introduction is achieved by solution nebulisation or with special techniques, like cold vapour method or hydride generation (Section 7.4).

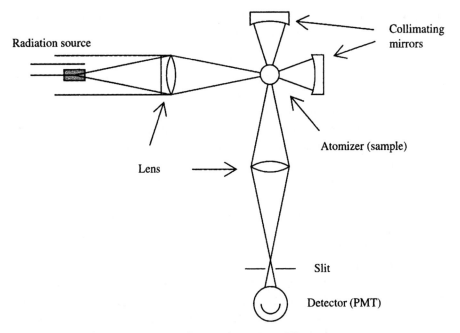

Figure 161 *Schematic structure of a non dispersive AFS instrument*
(Adapted from: P. L. Larkins, Spectrochim. Acta, 1971, **26B**, 477)

7.2.1 Radiation Sources

An ideal radiation source for AFS measurements should fulfill the following requirements: (*a*) high intensity, (*b*) good stability (short- and long-term), (*c*) ease of operation, (*d*) reasonable price, (*e*) available for all common elements, and (*f*) long operation time. The most used radiation sources in AFS are hollow cathode lamps, EDLs, ICPs, lasers, and xenon arc lamps.

7.2.1.1 Hollow Cathode Lamps. The intensities of the hollow cathode lamps normally used in AAS are too low for AFS determinations. However, the intensities may be increased when higher lamp currents are periodically used. Several hollow cathode lamps can be used at the same time for simultaneous multi-element analysis provided that the operations of the lamps and the monochromator or filters are synchronized. Special hollow cathode lamps (e.g. boosted discharge lamps) with high intensities have been developed. These lamps are especially well suited for AFS determination of hydride forming elements, such as As and Se, that are difficult to excitate.

7.2.1.2 EDLs. The intensities of electrodeless discharge lamps are better than those of normal hollow cathode lamps. In addition, the operation life of EDLs is long. However, the problem with these lamps is the long-term instability of the emission signal.

7.2.1.3 Lasers. Many features of lasers make them ideal radiation sources for AFS. The intensity of monochromatic and coherent radiation of the laser is very high, and the optical system employing the laser is simple. Disadvantages of lasers are their high price and limited wavelength range. The obtainable wavelength range is usually from 217 to 900 nm. Due to the high intensity of a laser, the dynamic linear range is often 10^5 or more.

When a laser radiation source is used, background correction can be carried out in several ways. For many elements measurements can be made using lines other than the resonance lines. Then the scatter at the excitation wavelength does not interfere. In this case, the detection limit is affected by the detector noise, background emission of the atomization cell, and molecular fluorescence. When using a resonance line, the detection limit obtainable is dependent on the light scatter of the radiation from the excitation source. The background correction can be carried out by separating fluorescence and background signals from each other with respect to time.

7.2.1.4 Plasma. In these methods the excitation radiation is achieved by aspirating pure solutions of the analyte element into the plasma. An ICP produces very intense and stable emission spectra with narrow spectral lines. The change of analyte elements from one to another is easy, only the solution introduced into the plasma is changed. For simultaneous multi-element analysis, a solution containing several analyte elements is employed.

When using an ICP for excitation and an air-acetylene flame shielded with nitrogen for atomization, the detection limits obtained for Ca, Cd, Co, Fe, Mg, Mn, and Zn were about the same or better than those obtained for the corresponding elements by flame AAS.

Due to the high excitation power of the ICP, the spectrum obtained is very rich with lines caused by atoms and ions, which makes it possible to correct the interference due to the light scatter using a double beam technique. When using a plasma source for excitation, interference caused by the scatter may be corrected by a method based on self-absorption. In this technique the slope of the fluorescence graph is compared to that of the plasma emission graph.

7.2.1.5 Xenon Arc Lamp. The xenon arc lamp emits a continuum spectrum which can be used for the determination of a number of elements. A xenon arc lamp of the 'Eimac'-type has been found to be best suited for AFS measurement. However, it is unstable when used periodically and its intensity decreases significantly at wavelengths below 210 mm. Thus, it is not suitable for the determination of arsenic or selenium. Detection limits obtained by a xenon arc lamp are about the same as those for flame AAS.

7.2.2 Atomizers

An ideal atomizer should have good atomization features and low background emission. In addition, the atomizer should not contain fluorescence quenching factors and the distance for atoms passing through the excitation zone should be as long as possible. It is quite difficult to unite all these requirements in one atomizer. Commonly used atomizers in AFS are flames, plasmas and electrothermal atomizers.

7.2.2.1 Flames. Various hydrogen flames have very small background emission and they do not contain fluorescence quenching particles. However, their atomization power is poor and several interferences may appear in these flames. They can be used for the determination of easily volatilized elements.

Air-acetylene and dinitrogen oxide-acetylene flames commonly used in flame AAS are also employed in AFS. However, the quenching of fluorescence radiation by combustion products (such as CO and CO_2) will reduce the sensitivity of the measurement. Nitrogen or argon have been used as shielding gases in order to reduce background emission from the flame. Most investigations have been carried out by using an argon-oxygen-acetylene flame or a helium-oxygen-acetylene flame. By altering the argon/oxygen (or helium/oxygen) ratio, optimum conditions (composition of the flame and temperature) for the determination of each element may be achieved. For example, easily volatilized elements are determined in a cool flame and elements which are difficult to atomize in a hot flame. This can be done without changing the flame- or burner-type.

Annular premix burners are better than direct injection burners, because the scatter of the light from the particles is smaller with premix burners. Scatter will be reduced especially when concentrated salt solutions are analysed.

7.2.2.2 Plasma. In 1976 Montaser and Fassel introduced an AFS instrument in which EDLs were used for excitation and an ICP for atomization. Advantages of the plasma torch are good atomization efficiency and lack of chemical interferences. Interference due to light scatter is also minimal with a plasma atomizer. In principle, it is also possible to analyse solid samples by AFS by using a plasma atomizer.

When an ICP is used for atomization, measurements cannot be made at the normal height above the coil (15–20 mm), since at this height the plasma emits strongly. In AFS longer plasma tubes are used and the viewing zone is about 45–65 mm above the coil.

The first commercial plasma atomic fluorescence spectrometer was developed by Demers and Allemand. Hollow cathode lamps are used as radiation sources and an inductively coupled plasma torch as an atomizer. Detection limits are reported for more than 30 elements. The linear dynamic range is normally 10^4 to 10^5.

7.2.2.3 Electrothermal Atomizers. In principle, electrothermal atomization is a very suitable method for AFS since, when an inert gas atmosphere is used, the fluorescence quenching will be minimized without losing atomization efficiency. On the other hand, the strong background emission of the hot graphite tube will decrease the signal/noise ratio.

In electrothermal AFS two techniques have been used for sample introduction: (i) the total sample is injected into the atomizer, or (ii) the sample as an aerosol is introduced continuously into the atomizer. In the first case, a multistep temperature programme is employed as in GF-AAS. In the latter case, the sample is first led to a spray chamber and then to a graphite furnace atomizer.

Several different electrothermal atomizers such as carbon rod, graphite ribbon, graphite furnace, and metal loop atomizers, have been designed for AFS measurements. In general, the electrothermal atomization method is time-consuming and expensive with respect to flame atomization. It is, thus, appropriate to use electrothermal atomization only when the sample amount or analyte concentration restrict the use of flames or plasmas.

7.2.2.4 Glow Discharge Chamber. This is used for the analysis of solid metal samples. Atomization is produced by sputtering and intensive excitation radiation is focused onto the atom vapour formed. The fluorescence radiation produced is then detected at right angles to the incident radiation. The precision obtained has been about 2% for many metals and the detection limits reported have varied between 1 and 100 mg

7.2.3 Selection of the Wavelength

7.2.3.1 Non-dispersive Instruments. In non-dispersive equipment, the atoms act as their own monochromators. In these instruments element-specific, line-like radiation sources (hollow cathode lamps, EDLs, lasers) are employed, and hence only the atoms of the analyte element are excited. For example, if a single element hollow cathode lamp or EDL is used, the fluorescence spectrum produced originates from the atoms of this particular element only. Then it is not necessary to isolate the different fluorescence lines from each other, and all the lines can be measured at the same time. The principle of non-dispersive AFS equipment was first introduced by Larkins.

The optimization of the signal/noise ratio is very important in AFS. In non-dispersive equipment, the intensity of the signal is directly proportional to the intensity of the incident beam, space angle, quantum yield, and transmission efficiency of the optics. The last factor may be improved by using mirrors because no colour deflections occur, unlike the case of lenses.

Advantages of the non-dispersive instruments are the simplicity and low costs. In addition, the optical transmission of the instruments is good, because monochromators are not required. In multi-element analysis, rotating filters may be employed to separate the wavelengths of different elements from each other. On the other hand, non-dispersive instruments are prone to interferences caused by stray light. In addition, background emission from the atomizer may cause interference. In order to minimize interferences flames shielded with nitrogen or argon are often used.

7.2.3.2 Dispersive Instruments. In dispersive instruments monochromators are employed for selection of the wavelength. When a line-like radiation source is employed, a monochromator of low resolution is adequate, but for a continuum radiation source a high resolution monochromator is required. In dispersive equipment the exit slit width is narrower than that in non-dispersive equipment. In this way, thermal background emission and stray light originating from the atomizer can be considerably decreased, but at the same time the optical transmission also decreases. The schematic construction of a dispersive AFS instrument is shown in Figure 162.

7.3 INTERFERENCE EFFECTS

In principle, similar interference effects are possible in AFS than in AAS and AES. Thus interferences due to differences in physical properties of samples and standards are possible when samples are introduced by solution nebulization. Chemical interferences owing to formation of stable compound may exist with flame and electrothermal atomizers, but are unlike in ICP-AFS.

Radiation
source

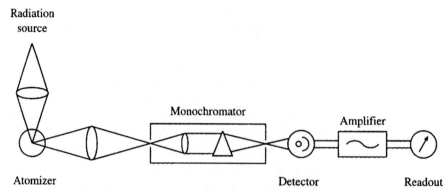

Figure 162   *Schematic structure of a dispersive AFS instrument*
(Adapted from: N. Ometto and J. D. Winefordner, Prog. Anal. Atom.
Spectrosc., 1979, **2**, 1)

Fluorescence radiation is usually measured at the right angle to the
incident radiation (Figure 161). Matrix interferences due to light scatter
from particles present in the atomizer, and molecular fluorescence
of the matrix compounds may appear. Matrix interferences may be
reduced by preparing sample-like standards. Matrix effects may also
be corrected by wavelength modulation and by using the double-beam
technique. Wavelength modulation is usually performed with a piece of
quartz or a filter.

The use of an ICP as an excitation source makes it possible to correct interference due to light scatter by means of several different
techniques. The double-beam technique is possible because ion and
resonance lines can be obtained by the ICP which do not appear in
cooler atomizers. In this technique, non-resonance lines near the
resonance line of the analyte are employed.

To avoid spectral interferences when non-dispersive AFS instruments are used, line-like radiation sources are required. Spectral interferences using narrow line sources are uncommon, while when using
continuum sources they are quite common. Corrections can be made
by wavelength or amplitude modulation, which can be performed with
filters. Different wavelengths are separated from each other by filters
whose advantage is their low price and good spectral transmission.

7.4   ANALYTICAL APPLICATIONS OF AFS

In principle, atomic fluorescence spectrometry is a very sensitive
analytical technique and a wide linear dynamic range should be
obtained. An ICP is a superior atomization source when compared

to combustion flames where matrix interferences and quenching of fluorescence radiation cause problems. However, in practice quite similar detection limits are obtained by HCL-ICP-AFS (excitation by boosted hollow cathode lamps) and ICP-AES. Thus ICP-AFS provides not much advantage over ICP-AES and commercial atomic fluorescence spectrometers instruments are currently not in market.

However, AFS is an important analytical technique in specific applications, *i.e.* when cold vapour method or hydride generation are used for sample introduction. In cold vapour method mercury is first reduced with tin(II) chloride to metallic mercury, and the mercury vapour is fed in an argon stream to the excitation cell.

Hydrogen flames have been employed in AFS for the atomization of gaseous hydrides. These flames are very suitable for hydride generation methods because of the low background emission. Hydrogen-air diffusion flame can be supported by the hydrogen generated as a side-product with the $NaBH_4$ reductant used.

Vapour generation is a very efficient sample introduction technique where the analyte is first separated from the sample matrix. Thus various interference effects during atomization are mostly avoided. The gas mixture entering the atomizer may be dried with a membrane dried shown in Figure 30 to avoid quenching of the fluorescence radiation. A mercury discharge lamp or boosted hollow cathode lamps are used for the excitation, and the fluorescence signal is detected at right angle to the incident radiation with a (solar blind) photomultiplier. Simple non-dispersive instruments, equipped with filters can be used.

Detection limits obtained by CV-AFS and HG-AFS are extremely low, below 0.01 p.p.b. for As, Bi, Sb, Se and Te and less than 0.001 p.p.b for Hg.

CHAPTER 8

Sample Preparation

For atomic absorption, plasma atomic emission, and atomic fluorescence spectrometric analysis the sample is normally introduced in the liquid form. The pretreatment of the sample depends on the sample type, element to be determined, and its concentration, as well as the analysis technique used. Any general sample preparation which would fit all kinds of samples cannot be employed. Sometimes the sample can be analysed without any pretreatment, sometimes complex sample treatment procedures are needed.

The chemical analysis consists of the following steps: (i) Sample collection and storage; (ii) Pretreatment and sample preparation; (iii) Measurement; (iv) Data processing and validation. Sample collection and storage are very critical steps in the whole procedure. Faults made during these steps cannot be corrected afterwards. Sample preparation must be taken into account in the calibration. The risk of contamination will increase with decreasing concentration of the analyte. Attention must be paid especially to contamination in the determination of ultra small amounts (p.p.b.-level).

In the following, the principles of some commonly used sample preparation methods for various atomic spectrometric analyses are detailed. For each element in a different sample matrices, detailed instructions for sample preparation can be found in the literature for different atomic spectrometric methods.

8.1 SAMPLE COLLECTION AND STORAGE

The sample collection method used depends on the sample type and amount. In practice, sample collection is done by someone other than the analyst. For example, blood samples are usually taken by hospital nurses and environmental samples by municipal authorities. In order to be sure that (i) the results of determinations bear comparison with other laboratories, (ii) solid conclusions can be drawn from the results, and (iii) the sample taken is representative, special attention must be paid to sampling, and, if possible, for various sample types confirmed sample collection instructions must be used.

Biological samples can be stored for short periods of time in a refrigerator and for longer periods of time in a freezer. Aqueous samples are always stored in plastic or teflon bottles. Immediately after the sample is taken, aqueous solutions are made acidic (pH 1–2) in order to minimize chemisorption and hydrolysis of metal ions. The precipitation of metal ions can also be prevented by addition of complexing agents.

Water content (moisture) of a solid sample is an important task. Many sample types contain substantial amounts of water. Samples may be dried before analysis or at least the water content of the sample must be accurately known. Often, if enough sample material is available, part of the sample is taken for moisture determination and parallel to that on portion is taken for analysis. After both determinations the final analysis result is reported on a dry weight or wet weight basis. In principle the moisture determination is an easy task; samples are dried at 105°C in a normal pressure in an oven. In practice, depending on the sample type, many difficulties may exist. Some elements and their compounds are easily volatilized and they may be partly evaporated even around 100°C. Mercury is the most common example. During drying, only water should be evaporated. However, especially biological samples are prone to changes if dried at elevated temperatures; organic and/or organometallic compounds may escape; the sample may be oxidized in the presence of ambient air, etc. Therefore samples should be dried at room temperature, for example in a dessicator, possibly under inert atmosphere or reduced pressure.

Many commercial apparatus are currently available for an automated sample drying. In these devices heating by infrared radiation or by microwave energy are employed. Freeze-drying (lyophilization) under reduced pressure is generally accepted as a safest way for drying biological samples.

8.2 SYSTEMATIC ERRORS: CONTAMINATION AND ADSORPTION

The risk of contamination increases as smaller analyte concentrations are determined. For example, graphite furnace atomic absorption and ICP mass spectrometry involve measurements of less than 1 ng of an element during the analytical procedure. At such low concentration levels, contamination of the instrument with detectable amounts of the elements to be determined is a severe problem. Contamination may occur at any stage of the procedure. The sample preparation should be as simple as possible since contamination may arise from the reagents and the vessels used during the preparation. In addition, the laboratory atmosphere may cause contamination at any stage in the procedure, even when the sample is situated in the atomization cell. Contamination problems are probable especially for the elements are common in earth´s crust (Si, Al, Fe, Ca, Na, K, Mg, Mn and Ti). Local emissions from industry may also occur (for example V, Sn, Cd, Ni, Cr and Ag). The instrument and sample preparation rooms should be separated from each other, and these should be so called clean-air rooms. In the case of GF-AAS and ICP-MS, the instrument rooms should be under positive air pressure with a dust filtration system (HEPA-filters). Sodium, magnesium, and zinc are common contaminants which are detected after conventional laboratory washing procedures. Zinc is particularly found in cleaning agents. Other frequently found contaminants are iron, copper, and potassium. The matrix may also be the potential contamination source, especially if the laboratory deals with a particular type of sample. The following precautions are essential in order to avoid contamination:

(i)   In every sample preparation step, the water used must be ultra pure with a minimum resistivity of 10 MΩ cm (the resistivity of pure water at 25°C is 18.18 MΩ cm). Several methods can be used for water purification, each having their advantages and disadvantages. These include distillation, deionization, reverse osmosis, ultrafiltration, membrane filtration and active carbon filtration. Nowadays commercial water purification systems are available all of which can use these techniques, often in several stages. The most pure water should not be stored, but prepared immediately before use. Organic compounds cannot be removed by conventional distillation, but removal by three successive distillations is possible. The water is first distilled from a basic permanganate solution, then from an acidic dichromate solution, and finally without additives. All charged particles may be removed by ion exchange, but not neutral ones like organic acids, amines, and silicic acid. These may be collected by active carbon filtration.

(ii) When determinations are normally carried out in aqueous acidic solutions, the acids used may cause contamination. Usually *pro analysis* grade acids are pure enough to use. When required, the acids used may be purified by distillation. Distillation of acids (and water) with a quartz still using infra-red heating (sub-boiling distillation) is used is for the preparation of the most pure reagents. Distillation is carried out in metal free clean area chamber.

(iii) All glass vessels to be used should be washed, rinsed, and then soaked in 10–20% (v/v) nitric acid for about 24 hours. Plastic vessels should be soaked in 1–5% (v/v) nitric acid. Finally, before use all vessels should be rinsed in high purity deionized water. Micropipette tips can also introduce contamination.

(iv) A clean 'bench' area is important to avoid contamination from dust and/or fume sources. Apparatus used in making solutions for trace elemental analysis should be kept separate from conventional laboratory work.

(v) Solutions of low concentration (about 20 mg l⁻¹ and below) should be prepared immediately before use, and it is not recommended to store them for long periods of time. Plastic containers of Teflon or FEP have been found to be most suitable for the storage of solutions.

(vi) An efficient fume extraction system should be fitted over each spectrometer.

Table 40 lists impurities in some commonly used laboratory vessel materials and reagents. In Table 41 impurities in some common acids and water are given. The impurity levels found in commercial reagents may vary considerably (Table 42). If needed, solid reagents may be further in-house purified, using a suitable method, such as electrolysis, sublimation, liquid-liquid extraction or chromatography.

Negative systematic errors may arise due to adsorption phenomenon (the opposite is also possible when adsorbed elements are released later due to inappropriate clean up of vessels). Especially when the element concentration is around 1 mg/l or below, adsorption processes may cause severe systematic errors. The adsorption processes are affected by (i) the materials and dimensions of used storage vessels and other equipment; (ii) clean up procedures used; (iii) the element itself (concentration, oxidation state); (iv) other elements present and their oxidation states, complexing agents present, pH value of sample solution; (v) storage time and temperature. To minimize the effect of adsorption phenomenon very dilute solutions should be measured immediately or stored minimum time in refrigerator. The concentration of test samples should be as high as possible and the surface area and volume of the equipment and vessels should be as low as possible.

Table 40 *Impurities ($\mu g\ kg^{-1}$) of some laboratory vessel materials*
(Adapted from: M. Stoeppler (Ed.), Sampling and sample preparation, Springer-Verlag, Berlin, 1997)

Element	PTFE	Quartz	Ultrapure quartz (Suprasil)	Laboratory glass (Borosilicate)
B		100	10	Main element
Na	25000	1000	10	Main element
Mg		10	100	6×10^5
Al		30000	100	Main element
Si		Main element	Main element	Main element
Ca		800–3000	100	10^6
Ti		800	100	3000
Cr	30	5	3	3000
Mn		10	10	6000
Fe	10	800	200	2×10^5
Co	2	1	1	100
Ni				2000
Cu	10	70	10	1000
Zn		50	200	3000
As		80	0.1	500–22000
Cd		10		1000
Sb	0.4	2	1	8000
Hg	10	1	1	

Table 41 *Impurities ($\mu g\ l^{-1}$) in suprapure water and some common mineral acids*
(Adapted from: M. Stoeppler (Ed.), Sampling and sample preparation, Springer-Verlag, Berlin, 1997)

			Cd	Cu	Fe	Al	Pb	Mg	Zn
H_2O		subboiling	0.01	0.04	0.32	<0.05	0.02	<0.02	<0.04
HCl	10 M	subboiling	0.01	0.07	0.6	0.07	<0.05	0.2	0.2
HCl	10 M	suprapure	0.03	0.2	11	0.8	0.13	0.5	0.3
HCl	12 M	p.a.	0.1	0.1	100	10	0.5	14	8.0
HNO_3	15 M	subboiling	0.001	0.25	0.2	<0.005	<0.002	0.15	0.04
HNO_3	15 M	suprapure	0.06	3.0	14	10	0.7	1.5	5.0
HNO_3	15 M	p.a	0.1	2.0	25	10	0.05	22	3.0
HF	54%	subboiling	0.01	0.5	1.2	2.0	0.5	1.5	1.0
HF	40%	suprapure	0.01	0.1	3.0	1.0	3.0	2.0	1.3
HF	54%	p.a.	0.06	2.0	100	5.0	4.0	3.0	5.0

Adsorption can be minimized by acidifying solutions to pH 1 with suprapure HCl or HNO_3. Complexing agents are also useful; for example Hg^{2+} is adsorbing onto a glass and PTFE, but HgI_4^{2-} is not. However, one should be aware of contamination when adding auxiliary reagents.

Table 42 *Impurities of some chelating reagents*
(Source: D. E. Robertson, Anal. Chem., 1968, **40**, 1067)

REAGENT	Ag	Co	Cr	Cu	Fe	Sb	Zn
Dithizone	<10	1.2	<2000	420	<7000	0.8	1150
Cupferron	3	0.68	<200	160	<600	0.3	7000
Nitroso-R	436	111	a	b	360000	40	1420
Thionalide	4.6	5.1	a	100	<300	3.7	120
2-Benzimidazolethiol	<2	21	a	0.4	69000	45	53000
1-Pyrrolidine carbodithioic acid	11	1.3	a	4	5000	1.9	1970
Sodium diethyldithiocarbamate	<10	0.56	a	b	<600	42	40
Thenoyltrifluoroacetone	209	4.9	192	16	11300	6.2	329
8-Hydroxyquinoline	<0.4	<0.4	a	b	1000	6.0	<100

[a]Not determined due to interference in the neutron activation analysis method used.
[b]Not determined.

Pure quartz and Teflon polymers are considered to be the best vessel materials for sample storage. Glass vessels should not be used due to contamination and adsorption problems if very low element levels are to be measured. The clean up procedure has also a marked effect onto surface structure of the vessel. Normal leaching procedures with pure acids and water should be avoided if extremely pure vessels are needed. Instead, steaming with HCl, HNO_3 and water is a very effective way to treat the vessels to minimize both contamination and adsorption problems.

8.3 SAMPLE PREPARATION METHODS

Samples may be divided into those which are already in a solution or liquid state and solid samples (particulate matter in the air, for instance, may be collected on filters using automated samplers). Solid samples may contain a high proportion of organic matter, or are entirely inorganic solids such as metals or rocks. In the sample preparation of organic materials (plant and animal tissue, blood, organic materials) dry ashing or wet digestion are the most common dissolution methods. In these methods organic sample matrix is oxidized to carbon dioxide and water and inorganic constituents are solubilized using mineral acids. Microwave digestion ovens are very convenient for wet digestion. Inorganic sample types are solubilized using wet digestion at ambient pressure or with pressure digestion using concentrated mineral acids and sometimes with fusion.

The important task that determines the sample preparation method is the purpose of the analysis, *i.e.* the result obtained after sample preparation and analysis must be valid for the purpose the analysis result was intended to use. Numerous sample preparation procedures have been developed to obtain answers to questions that could include: (i) What is the average gold concentration in the mineral deposit? (ii) What are the nutrient concentrations in soil samples that are available for plants? (iii) To which fraction element is associated in the soil sample (organic fraction, associated with Fe/Mn oxides, located inside a crystal lattice, etc); (iv) How much cadmium is likely to be released to the environment during 50 years if fly ash is used in road construction? (v) What is the concentration of extremely toxic methyl mercury (CH_3Hg^+) in a fish sample? It is easily understood that not a one sample preparation method can give answers to all the questions stated. Some aspects of sample preparation are discussed in the next paragraphs.

Traditionally, total concentrations of different elements in a variety of sample types are determined. Therefore the sample preparation method must be capable to decompose and dissolve all compounds where the analyte elements are associated with and bring the elements into solution for analysis. Determination of total concentration is often adequate. For example, determination of methylmercury in a fish sample is likely not to be necessary if the total mercury concentration in the sample is very low. When more information is needed, sample preparation methods (and analysis) are often more complex, needs more knowledge, time and laboratory skills. In addition, sophisticated and often expensive instrumentation is needed.

In geochemical survey aqua regia leach (HCl/HNO_3) of rock samples is generally suitable when one wants to find out the samples that are interesting and need to be studied in more detail. It can be generally assumed that aqua regia will dissolve almost completely most of the interesting elements present, despite the fact that the silicate matrix of the samples remains unattacked. The same is true for fly ash samples when the levels of toxic elements are determined in order to decide if the ash can be used in construction industry.

However, if one wants to determine total concentration of a particular element in geological sample the sample must be completely dissolved. This can be achieved by employing hydrofluoric acid in sample preparation to dissolve the silicate matrix for releasing the element/compound possibly entrapped in crystal lattice. Fusion procedures have to be sometimes employed for the most resistant sample matrices. Sampling and sample preparation is a very demanding task when one wants to know the average gold concentration in the mineral deposit studied, since very low gold levels are occurring and gold (as elemental form) is unevenly distributed as nuggets. In the sample preparation stage tenths of grams of grounded sample must be treated, e.g. with a fire-assay method.

Quite an opposite task to the determination of total concentrations is the determination of major nutrients in soil that are available for plants. There is little sense to employ aqua reqia leach in this connection since many minerals soluble to this mixture are not available for the plants. Instead, one wants to know the element fraction that is easily solubilized, *i.e.* the fraction that is adsorbed on soil particles. This can be approximated by shaking the soil sample with ammonium acetate (for example) to release the adsorbed particles by ion exchange.

In the case of contaminated soils and sediments and wastes, the determination of total concentrations of toxic metals gives an indication of the risk that may be caused to the environment. However, in addition to the determination of total concentrations it is also very important to estimate the mobility of toxic elements. Many chemical and physical characteristics of the contaminated site affect the mobility of a particular element present; these include pH value; content of organic matter; other nutrients present; cation exchange capacity, system redox potential, etc. These and many other variables affect the phases that may exist and how tightly different elements (depending on their chemical form(s)) are associated with these phases. Another important task affecting the mobility of metals is weathering; *i.e.* the effect of pH (e.g. acid rain), oxidation redox phenomenon, and temperature changes, and so on. In order to evaluate the characteristics of the sample itself and the environmental factors affecting the mobility of elements, numerous single step and serial extraction procedures have been developed.

Various single extractants are used in leaching tests, e.g. 1 M ammonium acetate or 0.1 M calcium chloride solution in evaluation of exchangeable fraction; 0.05 M EDTA solution (complexed fraction); 0.43 M acetic acid solution (fraction bound to carbonates); 0.5 M hydroxylammonium chloride at pH 1.5 (reducible fraction) and 8.8 M H_2O_2 and 1 M ammoniumacetate at pH 2 (oxidisable fraction).

Unfortunately, due to many different extraction procedures used the comparison of published results is often very difficult. During the last years a great deal of work has been carried out to harmonize many published procedures so, that the results obtained can be more easily compared with each other. For example, in the BCR protocol used in Europe, three-step extraction is applied for extraction of soil and sediment samples (Table 43). Often the residue obtained after step 3 is digested with aqua regia (step 4). The total concentration obtained after steps 1–4 should be close to the result obtained for the original sample with a direct aqua regia digestion.

The sample preparation in speciation analysis is often a demanding task. One has to be sure that the interesting species are quite completely

Table 43 *Three-step sequential extraction protocol (BCR)*

Step	Reagents	Released metals
1	0.11 M acetic acid, 40 ml/1 g sample	Water and acid soluble fraction (cation exchange sites, carbonates)
	— Shake for 16 hours and centrifuge	
2	0.1 M hydroksylammonium chloride (pH 2.0), 40 ml/1 g sample, or	Reducible fraction (iron and manganese oxyhydroksides)
	0.5 M hydroksylammonium chloride (pH 1.5), 40 ml/1 g sample (modified BCR procedure)	
	— Shake for 16 hours and centrifuge	
3	8.8 M H_2O_2 (10 ml, evaporate near dryness, repeat twice) + 1 M ammonium acetate (pH 2), 50 ml/1 g sample	Oxidisable fraction (bound to organic matter and sulphides)

released from the sample matrix during sample preparation, but at the same time the species conversion don't happen. Quite similar sample preparation methods to those in organic analytical chemistry are often employed. These include various solvent extraction procedures, e.g. supercritical fluid extraction or microwave assisted solvent extraction.

8.3.1 Liquid Samples

8.3.1.1 Aqueous Solutions. These samples can sometimes be introduced without any prior treatment. Among these are various water samples (natural, sea, and drinking water), wines, beers, and in some cases urine. It is of great help in the analysis, if the operator knows the approximate concentration of the analyte in the sample in order to decide whether or not some dilution is desirable. However, sometimes speciation problems may exist in analysis of aqueous solutions, *i.e.* the analyte element may exist in several chemical forms in the sample. These species may show different chemical properties and behaviour during analysis. For example organometallic compounds are often very volatile and may partly escape during thermal pretreatment in GF-AAS determination, although matrix modification is used. Hence calibration using aqueous inorganic standards solutions may result in systematic error. In this case sample digestion with acids that convert the different species to same chemical form is reasonable. Otherwise, in plasma techniques (e.g. ICP-MS), the measurement system (plasma) is quite insensitive although the analyte exists in different chemical forms (or even associated with colloid particles). However, in sample introduction interferences due to speciation may exist.

If the analyte concentration is high (more than about 200 times the reciprocal sensitivity), this may cause problems in AAS. Sensitivity in FAAS is reduced by removal of the impact bead, by burner rotation, or use of less sensitive absorption lines. In GF-AAS sample volume can be reduced or in some cases inert gas flow during atomization can be maintained. However, if these are not practicable, the solution should be diluted to bring it into the best range for measurement. On the other hand, if the concentration is too low, scale expansion or some chemical pretreatment or more efficient sample introduction techniques are required.

Sometimes it is necessary to add a spectroscopic buffer to aqueous samples in order to reduce interference effects. These buffers are normally used in concentrated form in order to avoid dilution of the samples.

8.3.1.2 Non-aqueous Solutions. In flame atomic absorption, non-aqueous samples can sometimes be run directly, but this depends significantly on the viscosity of the solution. The viscosity should be similar to that of water for which most nebulizers are designed. Only some organic liquids such as ethanol or methyl isobutyl ketone fulfill this condition and hence these are often used for dilution of organic liquids. Mineral oil and petroleum spirit, paints and drying oils, vegetable oils, and organic liquids require minimal pretreatment in FAAS analysis. Standards can be prepared in the pure solvent.

Sensitivity in flame atomic absorption is often significantly better in organic solutions than in aqueous solutions. Organic solvents are often used in separation and preconcentration methods (Section 8.4). In Table 44 sensitivities are compared for copper, cadmium, and antimony in various organic solvents with those in aqueous solutions measured by flame atomic absorption.

In FAAS an ideal organic solvent should possess the following characteristics: (i) Low viscosity; (ii) Good combustion characteristics (low background absorption and nontoxic combustion products); (iii) Low volatility; (iv) Immiscibility in water; (v) Not poisonous; (vi) Good extraction efficiency; (vii) Good nebulization efficiency; (viii) Easy to handle; (ix) Available in high purity form.

Volatile solvents (methanol, acetone, diethyl ether) will vaporize significantly before meeting the flame. The expanded vapour increases the flow rate of the gases in the mixing chamber by increasing the burning velocity and causes an erratic response. Aromatic compounds and halides are also unsuitable for spraying into the flames because they are incompletely combusted and produce smoky flames.

Table 44   *Enhancement values of the copper, cadmium, and anti-
mony signals in FAAS with various organic solvents*
(Source: J. C. Chambers and B. E. McClellan, Anal.
Chem., 1976, **48**, 2161)

Solvent	$A(org)/A(aq,)^a$		
	Cu	Cd	Sb
n-Butyl ether	2.32		
Butyraldehyde	2.09	1.77	0.62
Isopropyl acetate	2.00	1.44	
n-Butyl acetate	1.87	1.32	1.58
Toluene	1.83		
p-Xylene	1.80		
Methyl ethyl ketone	1.80		
2-Heptanone	1.77	1.35	1.05
Methyl isobutyl ketone	1.70	1.35	1.75
2-Octanone	1.58	1.07	1.50
Butyl butyrate	1.58	0.77	1.33
Ethyl acetoacetate	1.51	1.05	1.13
p-Dioxane	1.38		
2,4-Pentanedione	1.35	1.16	1.52
Cyclohexanone	1.29	1.05	1.36
1-Butanol	1.25	0.81	1.18
4-Methyl- 1,3-dioxalan-2-one	1.12	0.72	1.01
Methyl benzoate	1.09	0.44	0.68
Nitrobenzene	0.83		
1-Octanol	0.58		
1-Hexanol	0.45	0.65	0.87
Cyclohexanol	0.29		

[a]A(org): absorbance in organic phase; A(aq) absorbance in aqueous solution.

The most suitable solvents are C_6 or C_7 aliphatic esters or ketones and C_{10} alkanes. The most common organic solvent in FAAS is 4-methyl-2-pentanone (MIBK = methyl isobutyl ketone). Its nebulization, combustion, and volatilization properties are good for spraying into the flames. Other solvents commonly used are ethylacetate, butylacetate, and ethylpropionate.

In practice, organic solvents increase the amount of combustion compounds in the flame and, because of that, the ratio between the fuel and oxidant gases must always be optimized when organic solvents are sprayed directly into the flame. The flow rate of the fuel gas and the sample intake may be decreased or the flow of the oxidant increased.

Elements dissolved in organic solvents usually give the same response in graphite furnace atomization as the same elements in aqueous standard solutions. This is because the solvent is removed at the drying stage, and most organic metal complexes are converted to stable inorganic compounds at the thermal pretreatment stage. Thus, in principle any organic solvent can be used in sample preparation for GF-AAS, and the best solvent for a particular separation can be employed in solvent extraction. However, the only limitation is imposed by the difficulties in handling organic solvents with micropipettes. The pipetting precision or repeatability of non-aqueous solvents such as alcohols, ketones, and chlorinated hydrocarbons are usually much poorer (10% or worse) than that of aqueous solutions. Low boiling point solvents such as chloroform are difficult to handle. This problem can be avoided by careful choice of solvent, for example, chloroform can be replaced by 1, 2-dichloroethane which has a much lower vapour pressure.

In plasma atomic emission the number of organic solvents which can be sprayed directly into the atomizer is bigger than in FAAS (Table 45). Suitable dilutants are hexanol, diisobutyl ketone, diisopropyl ketone, acetic acid, 2-butoxyethanol, diethyl sulfoxide, carbon tetrachloride, xylene, nitrobenzene, aniline, pyridine, benzyl alcohol, and tributyl phosphate.

One problem often encountered is the build up of carbon residue on the tip of an injector tube, when organic solvents are introduced into the plasma. Therefore, small amount of oxygen is often added to the carrier gas stream to overcome this problem. Introduction of organic solvents having high vapor pressure will cause plasma overloading. This will have a varying effect on the response of the analytical lines, depending on their excitation properties. Plasma overloading can be reduced by various means, such as cooling the spray chamber or using cryogenic desolvation. Generally, solvents with low vapor pressure, such as xylene, are preferred. Kerosene is an even better alternative in this sense. With these solvents the analysis is uncomplicated, only a somewhat higher plasma power is needed compared to aqueous solutions.

Plasma emission spectrometry is often used in the determination of wear metals in used engine oils. This allows one to diagnose the condition of the engine; the increased levels of wear metals might predict failure of a particular engine part. Plasma technique is capable of atomizing tiny particles (~10 μm) suspended in the oil, but coarse particles must be separated, for example by centrifugation. Weighed samples are diluted by weighing a suitable amount of organic solvent (e.g. kerosene, 1:9, w/w). Calibration solutions are usually prepared by weighing commercial standard oil and diluting with base oil to the same mass as samples and then diluting by kerosene.

Table 45 *Introduction of different organic solvents into an inductively coupled plasma*
(Source: A. Miyazaki, A. Kimura, K. Bansho, and Y. Umezaki, Anal. Chim. Acta, 1982, **144**, 213)

Solvent	B.p (°C)	Vapour pressure at 20°C (mm Hg)	Ease of introduction[a]
Methanol	64.7	105	0
Ethanol	78.3	120	–
Propanol	97.5	15	+
Isopropanol	82.4	24	+
Butanol	117.5	5	+
Hexanol	157.9	1	++
Ethyl acetate	76.8	74	0
Butyl acetate	126.3		+
Isobutyl acetate	118	15	+
Amyl acetate	148.8		+
Isoamyl acetate	142	4	+
Acetone	56.3	175	0
MIBK	115.8	5	+
Diisopropyl ketone	125		++
Diisobutyl ketone	168		++
Acetylacetone	137.0		+
Acetic acid	117.8	12	++
Hexane	68.8	120	0
2-Butoxyethanol	171.2		++
Diethyl sulfoxide	189		++
Dioxan	101.4	30	0
Chloroform	61.2	105	–
Carbon tetrachloride	76.7	86	++
Benzene	80.1	76	0
Toluene	110.8	21	–
Xylene	140.6	4	++
Nitrobenzene	210.9	1	++
Aniline	184.6	1	++
Pyridine	115.5	16	++
Benzyl alcohol	205.4	1	++
Cyclohexane	80.8	82	0
Tributyl phosphate	180		++
Water	100	18	++

[a]++: very easy; +: easy; –: difficult; 0: impossible.

8.3.1.3 Emulsions. Introduction of organic solvents into flame or plasma is not without problems. High organic content may cause increased noise in the flame, or high solvent load may deteriorate the analytical performance of the ICP source. Also the organometallic standards will cause additional costs. Therefore it is advantageous if the organic material content of the samples can be diminished. Organic samples (for example oils) can be ashed and the residue dissolved in acids to obtain aqueous solutions. However, this time-consuming and there is a risk of contamination and element losses during sample preparation. Another approach is to prepare emulsified samples. The weighed oil sample is mixed with an emulsifier, such as ethoxy nonylphenol, or Triton-X-100. Sometimes tetralin is added beforehand to the oil. After addition the emulsifier the solution is mixed until a homogenous solution is obtained. Finally, water is added gradually to the final weight and the solution is mixed continuously to obtain a homogeneous emulsion sample. Calibration standards are prepared accordingly, using either organometallic standards or aqueous standard solutions. Preparation of emulsified samples is in principle a simple procedure. However, in practice the amounts of the components taken and the order how they are mixed are very important parameters. Also the used reagents and their amounts depend on the sample type analysed. Heating up the solution may help to obtain a homogenous emulsion.

8.3.2 Solid Samples

In most analytical techniques samples in liquid form are preferred, for instance due to ordinary sample introduction techniques used (e.g. solution nebulization in atomic spectrometry). Therefore analytes in solid samples must be first released from sample matrix before determination. An important point that must be taken into account at this stage is the scope of the analysis that defines the pretreatment procedure used. Term "digestion" is most often used in sample preparation when harsh reagents and conditions are used to release the analyte elements from the sample matrix. Otherwise, terms "extraction" or "leaching" are employed when analytes are more or less selectively released from sample matrix with the reagents used, often without complete dissolution of the original sample.

Sample digestion is affected by common reactions and equilibria: redox-reactions, acid-base reactions, complex formation, ion exchange, adsorption, etc. Hence good knowledge both the sample and analyte chemistry will certainly help to find suitable reagents and conditions to obtain reliable analytical results. Increased digestion temperature will generally improve digestion efficiency and speed up the whole procedure. Heating is achieved often by convection (e.g. hot plate), but nowadays more and more by microwave heating. In addition, digestion/ leaching are sometimes assisted by shaking, stirring or by using ultrasonic bath (ultrasonic-assisted extraction). The particle size of sample plays also an important role: finely grinded sample has a larger surface area and is attacked more effectively and rapidly.

Many types of solid organic samples (plant material, animal tissue, soils, food, feeding stuffs, organic chemicals, pharmaceutical products, *etc.*) are passed into aqueous solution after dry or wet oxidation of the organic material. Some organic chemicals and pharmaceutical products may also dissolve either in water or an organic solvent. Acid extraction may also be used for trace analysis in organic samples, for example nickel in fats and a number of trace metals in plant material. The sample is shaken with or stood overnight in contact with hydrochloric or nitric acid.

Wet digestion, fusion, and pressure digestion are common methods for the dissolution of metals, slags, ores, minerals, rocks, cements, and other inorganic materials and products. If the final solutions contain more than about 0.5% of dissolved material, the standards should also contain the major constituents in order to match the viscosity and surface tension.

8.3.2.1 Dry Ashing.   Dry oxidation or ashing may be used to remove organic material from samples. The sample is weighed into a suitable container such as a ceramic or metal crucible, heated at 400–700°C, and the residue dissolved in an appropriate acid (the residue may be taken up in almost any desired acid medium). The standards are prepared in the same acid medium as the sample finally appears.

The method is simple, and large sample series may be treated at the same time. The contamination risk is minimal provided that air in the laboratory is clean. This procedure also enables the concentration of trace elements.

However, the dry ashing method cannot be applied if volatile elements such as Hg, As, Pb, Sb, Se, Cd, and Mo are to be determined since they may volatilize during the ashing process. The degree of volatilization of the analyte is dependent on two factors: (i) What form the analyte is present in the sample, and (2) What is the chemical environment of the analyte. Hg, As, Pb, Sb, Se, Cd, and Mo may be volatilized easily during ashing. Bi, Cr, Fe, Ni, V, and Zn may also volatilize as metals, chlorides, or organometallic compounds. Oxidants are used as ashing aids in order to speed up the ashing and prevent the volatilization of the analytes. Commonly used ashing aids are magnesium nitrate, sulfuric acid, and nitric acid.

8.3.2.2 Wet Digestion.   Wet digestion is the most common method of dissolving solid samples and involves the use of a mineral acid or a mixture of mineral acids. Acid digestions are usually carried out either in glass or Teflon vessels. Teflon is less prone to contamination.

Contamination must be avoided, and the acids used must be at least analytical grade or supra pure.

Hydrochloric acid dissolves many inorganic salts (carbonates and phosphates) and some oxides. Aqua regia (HCl:HNO$_3$ (v/v) = 3:1) dissolves most metals including noble metals (Ag, Pt, Pd). Nitric acid may be used to oxidize organic material and metals passive to other acids. Hydrofluoric acid is used to decompose silicate in geological and tissue samples. However, volatile SiF$_4$ may be lost unless sealed vessels are used. Excess HF can be complexed with boric acid prior to the determination. Perchloric acid is a very strong oxidizing acid which is commonly used in conjunction with nitric and sulfuric acids to break up organic material. A mixture where HNO$_3$:HClO$_4$:H$_2$SO$_4$ (v/v) = 3:1:1 is suitable for a great variety of organic samples. In addition, there will always be oxidizing conditions in the solution which prevents the loss of volatile elements. However, the use of perchloric acid always involves an explosion risk and therefore solutions must not be evaporated to dryness. Sulfuric acid will prevent an explosion in solution because perchloric acid evaporates before sulfuric acid. The appearance of the white sulfuric acid fume indicates the complete evaporation of HClO$_4$ from the solution.

The wet digestion process may be speeded up, for example, by molybdenum(vi) or vanadium(v) ions. Other digestion aids used are hydrogen peroxide and potassium permanganate. H$_2$O$_2$ decomposes fats.

8.3.2.3 Digestion in Closed Vessels (Pressure Digestion). There are several drawbacks when samples are digested in open vessels. A key parameter determining the rate of chemical reactions (such as oxidation) is digestion temperature. Therefore a high temperature in sample digestion is desirable. However, at the normal ambient pressure the boiling points of common inorganic acids HCl and HNO$_3$ used often in sample preparation are around 110°C and 120°C, respectively. Therefore these digestion reagents are gradually distilled away when open vessels are used. Hence more reagents must be added during sample preparation to complete the digestion. Due to excess reagents needed and open vessels used, contamination might be a problem if the analyte concentration in the sample is low. Sample preparation is also time consuming.

Nitric acid alone is incapable to completely digest organic samples in open vessels. The temperature needed is usually at least around 150°C. However, complete digestion of aromatic structures need temperatures above 300°C! Therefore additional oxidizing reagents such as H$_2$O$_2$, are essential.

The unoxidised (often dissolved) residue in the sample solutions do not necessarily interfere in the atomic spectrometric techniques due to high temperature of the source (e.g. plasma). However, many analytical procedures suffer from interferences due to presence of different chemical forms of an analyte in a sample. For example, some organic selenium compounds are very difficult to digest. This may cause problems if hydride generation technique is used in selenium determination. It is possible that the difficult to digest compounds, although dissolved, do not form selenium hydride, resulting in chemical interference. Therefore high oxidation efficiency (low residual carbon content) in the sample digestion is desirable.

Pressure digestion is essentially a wet digestion procedure which is carried out in a closed vessel. The pressure inside closed vessel increases due to higher vapor pressure of the reagents at elevated temperatures and due to gaseous reaction products formed (CO_2 and H_2O and NO_x, if organic samples are digested with HNO_3). Advantages of pressure digestion are: (i) The digestion is much faster than conventional wet digestion; (ii) The procedure eliminates losses of volatile components (*e.g.* Hg or SiF_4); (iii) The decomposition of more difficult samples is possible. In this technique the sample is placed in a pressure decomposition vessel with a suitable mixture of acids (Figure 163). Typical acids used are nitric acid or a mixture of nitric acid and sulfuric acid. Despite the name of this technique, the higher attainable temperatures of the reaction mixture is the key parameter defining the efficiency of digestion.

Silicate rocks are frequently decomposed by this technique. The sample is moistened with aqua regia and some hydrofluoric acid is added. The vessel is sealed and heated for 30 minutes (200°C). After cooling some water and boric acid are added.

Biological samples have also been decomposed by pressure digestion. Typical acids for these samples are nitric or sulfuric. However, there is always a risk of explosion with organic material. For example, in a 25 ml vessel the maximum amount of organic material is 100 mg. At high temperatures and pressures the fat substances of the sample may form nitroglycerol with nitric acid.

A drawback of a conventional pressure digestion system is a long digestion time due to slow convective heating of digestion vessel(s). Therefore nowadays most sample digestions are carried out with microwave heating (microwave-assisted digestion). Samples are placed in specially designed microwave digestion vessels with a mixture of suitable acids and other reagents and the vessels are closed. The

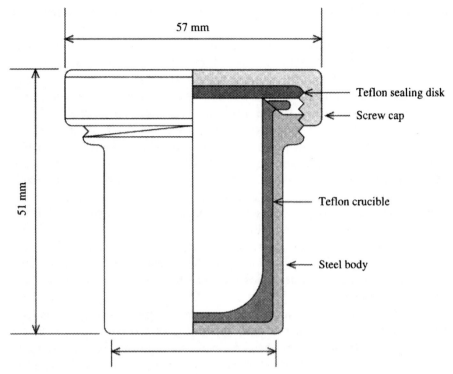

57 mm

51 mm

Teflon sealing disk

Screw cap

Teflon crucible

Steel body

Figure 163 *A pressure decomposition vessel*
(Adapted from: B. Bernas, Anal. Chem., 1968, **40**, 1682)

microwave digestion technique is based on high frequency microwaves (typically 2450 MHz). Polar molecules, such as water and mineral acids, absorb microwave energy. This results in rapid heating and temperatures of 100–250°C are reached within few minutes. Consequently, pressure is increased inside the closed vessels. The advantages of microwave-assisted digestion are low contamination (small amounts of pure reagents in closed vessels) and short digestion times. For example, a 0.2 g biological sample may be digested in 10–15 minutes, but the vessels must be left for half an hour to cool down before opening.

Microwave digestion systems are termed from low pressure systems (pressure inside vessels below 10 bar) to medium and high pressure systems (pressures up to 80 bar and higher). Vessels are equipped with pressure relief valves to avoid explosion. Various types Teflon polymers, such as PFA, are used for preparation of digestion vessels. Teflon is an ideal material since it is very pure and chemically resistant. To obtain the lowest contamination level, inserts made of pure quartz can be used inside the Teflon vessels. Drawback of an ordinary microwave digestion is the maximum available temperature that is limited to 250 °C due to heat softening of Teflon.

Many experimental variables affect the digestion efficiency with microwave-assisted digestion: type of microwave oven and digestion vessels, number of vessels, sample type and mass, digestion reagents used and their amounts, etc. Earlier, when digestion programmes were documented using power and time settings, the reproducibility of the microwave digestion procedures was often poor. Currently the internal temperature and pressure of the vessels are monitored during digestion. Since sample temperature is the most important parameter affecting the digestion reactions, the microwave oven programmes used for different sample types are documented on time/temperature basis. The temperature inside the vessel is raised to a desired level, say 180°C, during few minutes and kept there typically for 10 min with the aid of temperature control. The internal pressure of the vessel is monitored as a diagnostic value only. Thus, if the internal pressure rises too rapidly, the feedback system slows down the reaction by reducing the microwave power.

One drawback of ordinary microwave sample preparation is the limited sample mass (1–2 g or less, depending on the content of organic matter). In a focused open vessel microwave digestion system samples are digested in quartz or glass vessels. The lower part of vessel is heated by focused microwaves. By the use of sulfuric acid very high digestion temperatures are obtained and large amounts (10 g) of difficult to digest samples (such as polymers) can be digested. Due to open vessels the pressure is not a problem since the gases released during digestion are extracted. Microwave heated autoclave is also available for the digestion of very difficult samples. The autoclave is pressurized with high-purity nitrogen (up to 200 bar). Through microwave heating sample temperatures up to 350°C can be reached. At such a high temperature nitric acid, for example, is a very powerful oxidant.

8.3.2.4 Fusion. Difficult geological and metallurgical samples may be dissolved by this technique. A weighed sample is mixed with a suitable flux or a mixture of fluxes in a metal or graphite crucible. The sample and flux mixture is then heated over a flame or in a furnace and the resulting fusion cake or melt is dissolved either in water, suitable acid, or alkali. A number of different fluxes can be used and some of these are sample specific.

Sodium peroxide is a universal flux and fusions with this material are usually carried out in Zr crucibles. The cooled melt is leached in water or dilute acid solution. Lithium metaborate is a frequently used flux for silicates and fusions are carried out in Pt crucibles. The melt is leached with dilute acid solution.

Sodium carbonate fusion has been used to decompose silicate, oxide, phosphate, and sulphate minerals. A Pt crucible is usually used to carry out the fusion and the cooled melt is leached with a dilute acid solution.

Sodium tetraborate or combinations of boric acid, boric oxide, or sodium tetraborate with sodium carbonate have been used to decompose aluminosilicates and refractory minerals such as zirconium or chromium-bearing ores. Pt crucibles are used and dilute hydrochloric acid is used for leaching the melt.

Fusions are often the easiest and most successful technique used to decompose complicated samples, but it is not the first method of choice for determinations of ultra small analyte concentrations. The main problems involved are contamination from the flux and the high salt content of the resulting solution. Easily volatile elements may also escape during fusion.

8.3.2.5 Slurries. Atomic spectroscopic techniques are capable of atomizing solid particles suspended in liquid. These slurries may be introduced into atomizer like conventional liquid samples. In slurry technique the sample is grinded to a fine pulver (particle size should be uniform and below 10 µm). The grinded sample is mixed with solvent (often water). To prevent the settling of particles during the measurement, a stabilizer is often added (e.g. Triton X-100®). In addition, mixing with stirring, or preferably with ultrasonic agitation is needed. Use of slurry technique is advantageous when the dissolution of sample is difficult or time-consuming. Although in principle very simple technique, contamination during sample pulverization may cause problems. Finding suitable calibration standards may also be difficult.

8.4 SEPARATION AND PRECONCENTRATION METHODS

In trace elemental analysis, the analyte element must often be separated from an interfering matrix. Due to insufficient sensitivity of the analytical technique, attempts are often made to bring the analyte concentration in the sample solution within the measurable range by using suitable preconcentration (enrichment) methods. However, every additional sample pretreatment step is a potential source of error (*e.g.* contamination). Enrichment and separation steps are also time-consuming procedures. Thus, these steps should be avoided if possible.

For atomic spectrometric methods the non-specific separation of several elements is usually adequate. Separation and enrichment methods used for many years are liquid-liquid extraction, ion exchange, coprecipitation, and evaporation. Solid phase extraction using various supporting materials is coming increasingly popular also in the field of inorganic analytical chemistry.

8.4.1 Liquid-liquid Extraction

Liquid-liquid or solvent extraction is a procedure in which two immiscible liquids are shaken with each other so that one or more elements in one liquid phase are transferred to the other. One liquid phase is usually the aqueous sample solution and the other one is an organic solvent. Using solvent extraction, the analyte will be separated from concomitants, and an increase in the concentration is accompanied by a decrease in interfering influences.

8.4.1.1 Extraction Systems. In order to get metal cations transferred from an aqueous phase to an organic phase they must be converted from the ionic form to a neutral compound. In atomic spectrometric methods two extraction types are normally used: (i) chelate extraction and (ii) ion-pair extraction. These extraction processes can be represented by the following equations:

$$M^{2+}(aq) + \begin{pmatrix} {}^-L \\ {}^-L \end{pmatrix} (aq) \rightleftharpoons M \begin{pmatrix} L \\ L \end{pmatrix} (aq) \rightleftharpoons M \begin{pmatrix} L \\ L \end{pmatrix} (org) \qquad (84)$$

a bidentate   a neutral metal
ligand anion   chelate

$$M^{2+} (aq) + (L\text{-}L) (aq) \rightleftharpoons M(L\text{-}L)^{2+} (aq) \qquad (85)$$

a neutral   a cationic metal
bidentate   chelate
ligand

$$M(L\text{-}L)^{2+} (aq) + 2X^- (aq) \rightleftharpoons M(L\text{-}L)X_2 (aq) \rightleftharpoons M(L\text{-}L)X_2 (org) \qquad (86)$$

a neutral ion-pair

$$M^{2+} (aq) + 2(L\text{-}L)^{2-} (aq) \rightleftharpoons M(L\text{-}L)_2^{2-} (aq) \qquad (87)$$

an anion metal chelate

$$M(L\text{-}L)_2^{2-} (aq) + 2Y^+(aq) \rightleftharpoons Y_2M(L\text{-}L)_2 (aq) \rightleftharpoons Y_2M(L\text{-}L)_2 (org) \qquad (88)$$

a neutral ion-pair

In addition to these reactions, metal ions and ligands may participate in many other equilibria reactions such as complex formation, protolysis, and hydrolysis reactions. In order to transfer metal ions quantitatively into the organic phase, the extraction conditions (pH, reagent concentrations, volume ratio of the liquid phases) must be optimized. Figure 164 represents the response of molybdenum as a function of pH in three different organic solvents with toluene-3,4-dithiol and 8-hydroxyquinoline as ligands. According to this plot the determination of molybdenum is most sensitive in MIBK (methyl isobutyl ketone) for

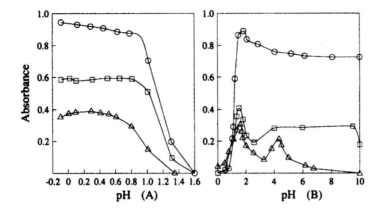

Figure 164 *Extraction of molybdenum as a function of pH. A: the Mo-dithiol system; B: the Mo-8-hydroxyquinoline system; (○) MIBK; (□) DIBK; (△) isoamyl alcohol*
(Adapted from: L. H. J. Lajunen and A. Kubin, Talanta, 1986, **33**, 265)

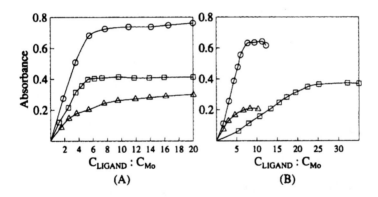

Figure 165 *The dependence of the degree of the extraction on the ligand/molybdenum molar ratio for various extraction systems. A: the Mo-dithiol system; B: the Mo-8-hydroxyquinoline system; (○) MIBK; (□) DIBK; (△) isoamyl alcohol*
(Adapted from: L. H. J. Lajunen and A. Kubin, Talanta, 1986, **33**, 265)

both ligands. The optimum pH range for 8-hydroquinoline and dithiol are from 3 to 10 and from –0.1 to 0.9, respectively. Figure 165 in turn shows the Mo response as a function of the ligand/metal concentration ratio. In the case of dithiol the ratio must be at least 5, and in the case of hydroxyquinoline at least 20.

Table 46   *Some common chelating agents used in liquid-liquid extraction*

Ligand	Abbreviation or trivial name	Structure
Ammonium pyrrolidine dithiocarbamate	*APDC*	
Sodium diethyldithiocarbamate	*NaDDC*	
8-Hydroxyquinoline	*Oxine*	
1,5-Diphenylthiocarbazone	*Dithizon*	
2,4-Pentanedione (acetylacetone)	*acac*	
Ammonium N-hydroxy-N-nitrosobenzeneamine	*cupferron*	

Table 46 lists some commonly used ligands in chelate extraction. All these ligands are polyprotic acids which mean that the metal complex formation with these ligands depends on the pH of the solution.

The most universal liquid-liquid extraction system in atomic absorption is the combination of APDC (ammonium pyrrolidine dithiocarbamate) as the ligand and MIBK (methyl isobutyl ketone) as the organic solvent. APDC forms stable metal complexes with a number of metal ions over a wide pH range. In addition, MIBK exhibits ideal combustion properties in flame atomic absorption. Its solubility in water is relatively high (2.15 ml per 100 ml H_2O at 30°C). In addition, its solubility significantly depends on the temperature and the ionic strength of the aqueous phase. The solubility of MIBK in water may be reduced and extraction efficiency increased with addition of amyl acetate (1:10 v/v) or cyclohexane (1:4 v/v). In Table 47 the optimum pH ranges are given for a number of metal ions when extracted as APDC chelates into MIBK. Usually 1–2% aqueous solutions of APDC are used. This reagent solution cannot be stored and it should be prepared just before use.

Selectivity of solvent extraction may be improved by using masking agents such as fluoride or cyanide ions in connection with 8-hydroxyquinoline. 8-Hydroxyquinoline reagent solution is prepared in an organic solvent (chloroform, butanol, ethylacetate, MIBK).

Dithizon is dissolved in chloroform and used as 0.002–0.2% solutions. Sodium diethyldithiocarbamate (NaDDC) and cupferron are used as 1-5% aqueous solutions.

In ion-pair extraction frequently used anions are halides, thiocyanide, and tributylphosphate. Ion-pairs are extracted from relatively acidic aqueous solutions to relatively polar organic solvents such as alcohols, ethers, esters, and ketones (Table 48).

The ion-pair extraction is often more effective than chelate extraction for preconcentration of the analyte element. Ion-pair extraction may also be used to separate the interfering matrix components from the analyte. In this case, the analyte will stay in the aqueous phase.

In synergic extraction, two ligands are used. One of these neutralizes the positive charge of the metal ion, and the other one is a neutral ligand. The neutral ligand also binds the metal ion by changing the physical and chemical characteristics of the complex compound. The

Table 47 *Optimum pH ranges for various metal ions using the APDC-MIBK extraction system*
(Source: G. F. Kirkbright and M. Sargant, 'Atomic Absorption and Fluorescence Spectroscopy', Academic Press Inc., New York, 1979)

pH range	Metal
2	W
2–4	Nb, U
2–6	As, Cr, Mo, Te, V
2–8	Sn
2–9	Sb, Se
2–14	Ag, Au, Bi, Cd, Co, Cu, Fe, Hg, Ir, Mn, Ni, Os, Pb, Pd, Pt, Rh, Ru, Tl, Zn

Table 48   *Ion association-ion pair extraction systems*
             (Source: G. H. Morrison and H. Freiser, Solvent Extraction in
             Analytical Chemistry, John Wiley, New York, 1957)

Metal ion	Aqueous phase	Organic solvent
Fe^{III}	1 M HCl	0.1 M TOPO[a]-cyclohexene
Fe^{III}	4 M HBr	Diethylether
Fe^{III}	6 M HCl	Diethylether
Fe^{III}	8 M HCl	Isopropylether
Sc^{III}	8 M HCl	Tributylphosphate
Ti^{IV}	7 M HCl	0.1 M TOPO-cyclohexene
V^{V}	pH: 1.5–2.0	0.6 M TOPO-kerosene
Zn^{II}	1 M NH$_4$SCN+ 0.5 M HCl	Diethylether

[a]TOPO = trioctylphosphine oxide.

mixed ligand complex formation increases the extraction of the analyte into the organic phase. For example, copper can be extracted with salicylic acid (H_2L) and pyridine (py). The mixed ligand complex formed is $Cu(HL)_2py_2$. Pyridine considerably improves the extraction. The extraction percentage for copper with salicylic acid alone is only about 5%, whereas that together with pyridine is over 30%. Vanadium may be extracted into dibutylether as VOL_2BuOH mixed ligand complex (where L = isopropyltropolone and BuOH = 1-butanol).

8.4.1.2   Standards and Blank Samples.   When extraction systems are used, it must be taken into account that the analyte is not always transferred completely into the organic phase. The extraction efficiency depends on the conditions such as pH of the aqueous phase and metal-ligand ratio. In addition to the organic solvent used, the ligand may also have an influence on the sensitivity of the analyte in FAAS. Thus, the sensitivity of the analyte may differ in the same solvent when different ligands are used, since the metal-ligand bonds in metal complexes formed possess different thermal properties. For example, the Mn sensitivity in MIBK is better with the cupferron complex than with the diethyldithiocarbamato or 8-hydroxyquinolato complexes.

For the reasons above, samples, standards, and blanks must be extracted in exactly the same way. In addition, standards and blanks must be pretreated, as much as possible, in the same way as the samples in order to obtain matching matrices.

If the extraction is not quantitative, several successive extractions are needed and the organic extracts are then combined. The disadvantage of this procedure is the diminished analyte concentration in the increased solution volume. However, organic phases may be easily concentrated by evaporation.

8.4.1.3 Back Extraction. The analyte may be back extracted into the aqueous phase with an acidified aqueous solution. The pH of the aqueous solution must be low enough to ensure the quantitative decomposition of the metal complex. The pH depends on the stability of the metal complex:

$$M(L-L)_2(org) + 2H^+ \rightleftharpoons M^{2+}(aq) + HL-LH \ (org/aq) \tag{89}$$

In this way it is possible to obtain complete separation of the analyte from the matrix components. The determination is then carried out in an almost pure aqueous solution of the analyte that exhibits hardly any interferences.

Back extraction is desirable when the organic solvent used cannot be introduced directly into the atomizer (flame, graphite furnace, or plasma). In electrothermal atomization, organic solvents are spread within the graphite tube and because of this the sensitivity is often better for aqueous solutions. The reproducibility of the determination is usually worse in organic solvents than in aqueous solutions.

8.4.2 Ion Exchange

Ion exchange is used when elements which are difficult to separate from each other (lanthanides, actinides) are to be determined. Ion exchange is a highly selective separation method and it is not often required. It is used in AAS and AES for group separation and removal of interfering concomitants or removal of an excess of a particular matrix ion. For example, boron may be determined by DCP-AES in steel by using sodium carbonate fusion for sample preparation. The excess of sodium coming from the fusion process will interfere with the measurements, but it can be removed by cation exchange. It is not necessary to pass the sample solution through a cation exchange column, stirring the regenerated cation exchange resin with the sample solution in a beaker is quite enough.

8.4.3 Solid Phase Extraction

Although liquid-liquid extraction is a simple and reliable method, large amounts of pure, often toxic organic solvents are needed. During last twenty years solid phase extraction has gained wider acceptance as a separation and enrichment technique in the field of inorganic analytical chemistry. In solid phase extraction (as in liquid-liquid extraction) the dissolved analyte elements (present as simple inorganic ions or organic complexes) are partitioned between two immiscible phases, liquid phase and solid phase. In inorganic analysis the liquid phase is often aqueous solution. To be effective, the distribution ratio between the solid phase and aqueous phase must be high enough. Various mechanisms prevail, depending on the sorbent and chemical form of the metal. These include adsorption, complex formation and ion exchange. Also ion pair formed with metal ions with a large oppositely charged organic counter ion is retained by solid sorbents.

In short, solid phase extraction is carried out as follows: (i) The solid phase is conditioned (activated) with a suitable solvent (e.g. methanol, if reversed-phase sorbent is used); (ii) Sample solution is put through the solid phase, e.g with a syringe; (iii) Interfering matrix components may be eluted with a weak eluent so, that the analyte elements are retained on the sorbent; (iv) Analyte elements are eluted with a suitable solvent. The elution must be quantitative with minimum amount of eluent to avoid dilution of a sample.

Various types of support materials are used as solid phases. These may be divided to inorganic and organic materials. The most common inorganic material is silica gel, SiO_2. Other oxides employed are Al_2O_3, TiO_2, ZrO_2 (amphoteric oxides) and MgO (basic). The affinity of these materials toward anions and cations depend on solution pH value. Generally, the selectivity of these materials is poor.

More selectivity is obtained when the support material is modifier with a suitable reagent. This can be achieved by chemical bonding (e.g. octadecylbonded C_{18} silica gel) or loading the adsorbent with the reagent solution (e.g. complex forming agent). The modified materials have a higher affinity towards non polar molecules. Therefore metal complexes or ion associates are passed through the adsorbent in separation process. For example, Cr^{VI} is retained by C_{18} phase under acidic conditions, when DDTC is used as a complexing agent. Octadecyl-bonded silica can be further modified by a suitable complexing agent to obtain better selectivity for particular ion(s).

Organic support materials used in solid phase extraction are numerous. These include synthetic materials (polystyrene-divinylbenzene copolymers, polyacrylate polymers, polyurethane foams, activated carbon, etc.) or natural materials (cellulose, chitine). These materials can be used for adsorption of metal ions or complexed metal ions. Also these materials can further modifier to obtain higher affinity for particular ions (e.g. a chelating resin containing iminodiacetate).

The applications of solid phase extraction in inorganic analytical chemistry are very wide in scope. In addition to analyte preconcentration and matrix separation, solid phase extraction can be applied during sampling, for example in separation of (unstable) organometallic species in water samples. Minicolumns filled with a suitable sorbent can be used together with FI technique to obtain a fully automated sample handling and determination.

8.4.4 Precipitation

In trace elemental analyses the analyte element may be separated from the concomitants by coprecipitation. Direct precipitation is not possible because of the small concentrations of the analyte in the sample solutions. However, sometimes direct precipitation can be used for the removal of matrix elements.

Coprecipitation always occurs in connection with the precipitation of sparingly soluble compounds. Foreign ions to the insoluble compound are usually adsorbed on the surface of the precipitate. Coprecipitation is quantitative only for very small concentrations.

If the sample contains plenty of inorganic cations, the trace components of the solutions can be separated by precipitation of the main components with a suitable reagent or by increasing the pH of the solution. However, some suitable metal cation must usually be added into the solution to obtain sufficient precipitation. Commonly used cations are Fe^{III}, Al^{III}, trivalent lanthanide, Mn^{II}, and Mg^{II} ions. Compounds to be precipitated are often hydroxides, halides, sulfides, or sulfates. The cation used must not cause any interference in the determination, and it must be added in sufficient quantity to ensure adequate precipitation.

Organic compounds such as 1-nitroso-2-naphthol, thionalide, 2-mercapto-benzimidazole, and potassium thiocyanate have also been used as coprecipitation reagents. These compounds, except KSCN, are insoluble in water and form insoluble complex compounds with a number of metal ions. Precipitation reagents are first dissolved in small amounts of volatile water insoluble organic solvents. These solutions are then added to the sample solutions and the organic solvents are evaporated. The precipitates obtained are dissolved in suitable organic solvents from which the determinations are carried out.

8.4.5 Evaporation

Evaporation is a simple, but slow concentration method. For solutions with high salt concentrations it is not desirable. Evaporation is useful a preconcentration method in special cases. For example very clean snow or ice samples can be melted and handled in clean room; every effort is taken to avoid contamination. The concentrated sample has a very simple matrix for further treatment in the analytical procedure.

CHAPTER 9

Hypheneted Techniques

The determination and identification of traces of inorganic, organic, and organometallic compounds in varying type of samples (environmental, industrial, biological and pharmaceutical samples) is of great value for many branches of science. The toxicity of a compound *in situ* depends on its exact nature, and hence speciation of metals and metal compounds is extremely important.

With an ordinary sample introduction techniques employed in atomic spectrometric techniques and ICP-MS (usually solution nebulisation), the signal depends on the total concentration of a particular element in a sample solution, despite the possible different chemical forms (species) existing in the sample solution (or otherwise interference exists). Therefore a suitable separation method must be used if different species are to be determined. There are number of publications that deal with speciation studies in limited special cases. For example Cr^{6+} (toxic species) can be separated and determined in the presence of Cr^{3+} by selective chelatation. For instance sodium diethyldithiocarbamate can be used for complexing Cr^{6+}. The chelate formed can be separated by solid phase extraction (Section 8.4.3) for the determination of Cr^{6+}. In an other case Cr^{3+} and Cr^{6+} can be in situ separated in GF-AAS determination using thenoyltrifluoroacetonate as a complexing agent. Cr^{3+} forms a volatile complex that is evaporated during ashing phase allowing a separate determination of Cr^{6+}. These kinds of determinations can be easily automated.

However, when several species (and different elements) are to be determined, on-line separations are needed. Separation methods most often employed are gas or liquid chromatography and sometimes capillary electrophoresis. Hyphenated analysis methods with element specific detection provide useful qualitative and quantitative information. In addition, interferences caused by overlapping chromatographic bands may be minimized by selective detection. Various GC and HPLC applications involve, for example, the determination of proteins, environmental toxins (methyl mercury), and pesticides.

Suitable sample collection/extraction procedures must be employed in sample pretreatment, depending on the sample type (gas, liquid or solid). The extraction procedure must release the species to be determined as completely as possible from the sample matrix, without affecting the chemical forms of different species. In addition, the sample preparation method must be compatible with the analysis method used; e.g. in gas chromatography the analytes must usually be transferred into organic solvent before analysis. However, with HPLC aqueous solutions can be employed.

9.1 ATOMIC SPECTROMETRIC DETECTORS

Atomic spectrometric techniques (AAS, AES, AFS) and ICP-MS provide a highly specific and sensitive analytical method for a variety of metal species. However, the efficiency of sample introduction technique employed is a very important parameter that has a great effect on the detection limits obtained in practice.

Atomic absorption spectrometry is available in most laboratories and can be employed with different atomization modes. Flame atomization is a very useful atomization technique since many kinds of solvents can be introduced, when coupled with HPLC. However, due to poor sample introduction efficiency with solution nebulisation and the short residence time of analyte atoms in the flame result in too high detection limits in many practical applications. The use of slotted quartz tube may help in some instances.

An electrically (or flame heated) quartz tube (QT) can be used for atomization of volatile organometallic compounds or their derivatives. A good sensitivity is obtained due to long residence times of analyte atoms. Also the path of the radiation passing through the atomic vapour is quite long which also improves the sensitivity. The carrier gas flow cools down the quartz tube and affects the residence time of analyte vapour. The optimum atomization temperature depends on the species analyzed (Figure 166).

Atomic absorption spectrometry with graphite furnace atomization is a very sensitive technique. However, the discontinuous functioning of the atomizer causes difficulties when coupled with chromatographic separation techniques.

Atomic fluorescence in turn provides a very sensitive detection method for the determination of mercury and hydride forming species.

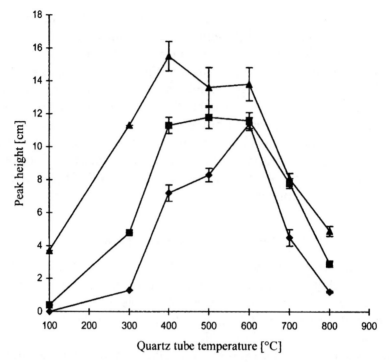

Figure 166   *Effect of the temperature of the electrically heated quartz tube on the detector response for different mercury species (about 5 ng of Hg in each case) determined by GC-QT-AAS.* ◆ *dimethylmercury,* ■ *methylethyl mercury,* ▲ *diethylmercury*
(Adapted from: I. Välimäki and P. Perämäki, Microchim. Acta, 2001, **137,** 191)

In GC-AFS volatile mercury species are atomized in electrically heated quartz tube. Different species of hydride forming elements may be separated with HPLC. The species that don't form hydrides with NaBH₄ must be oxidized to simpler species before hydride generation reaction.

Various plasma emission techniques employing microwave plasmas, direct current plasmas and inductively coupled plasmas have also been employed in many speciation studies. McCormack with his co-workers first interfaced a gas chromatograph to a microwave plasma in 1965. After that, numerous publications appeared describing the applications of different plasmas to gas or liquid chromatographic detection. In addition to metallic elements, hydrogen, carbon, nitrogen, phosphorus, oxygen, sulfur, halogens, and organic compounds have all been determined by these techniques. Microwave plasmas can be generally used only together with gas chromatography since the introduction of liquids by solution nebulisation generally extinguishes the plasma. Commercial atomic emission detectors employing MIPs are supplied by the manufacturers of gas chromatographs.

DCP-AES and ICP-AES can be directly coupled with liquid chromatography. However, sample introduction efficiency with solution nebulisation is low and the sensitivity obtained may not be sufficient for all applications.

ICP-MS is inevitably best suited detection technique for speciation studies. It is a very sensitive technique that can be quite easily coupled with various separation methods. The sensitivity with solution nebulisation is most often good enough to monitor the components in the HPLC mobile phase in real time. In practice simultaneous analysis is possible with conventional quadrupole ICP-MS instruments.

9.2 SEPARATION METHODS

The analytes pass into the element specific detector through chromatographic separation as hydrides or other volatile compounds, metal chelates, anions, or organometallic compounds. When the exact composition of the metal complex studied is known, it is possible to calculate the amount of the organic ligands on the basis of the amount of metal in the sample. Detection limits obtained by hyphenated methods are generally somewhat higher than those for conventional methods, but the sample volume required for the sample introduction by HPLC or GC is much smaller than that required for the solution nebulization.

The coupling of column outlet and the detector is usually easily achieved. In liquid chromatography the effluent can directly enter the nebulizer of the instrument. In some cases derivatization reaction, e.g. hydride generation, may be carried out. In gas chromatography the nebulizer must be removed and, for example in ICP-AES, an additional plasma gas flow must be employed. The transfer line between the column and atomizer should be as short as possible (and the dead volume should be minimal) to avoid peak broadening and worsening of resolution. In GC the transfer line must be heated to avoid condensation of gaseous species.

9.2.1 Gas Chromatography

Volatile compounds can be separated with high resolution using this technique. Currently capillary columns with dimethylpolysiloxane stationary phases are most often used in the separation of nonpolar species or their derivatives. Some species of trace elements are volatile enough for direct analysis after separation from sample matrix. These include $(CH_3)_2Se$ and $(CH_3)_2Se_2$ and organic mercury compounds CH_3Hg^+ and $(CH_3)_2Hg$. The volatile species can be separated after sample preparation with extraction using non-polar solvent and introduced to a gas chromatograph. However, most inorganic species are polar compounds with high boiling points and can not be directly analyzed by GC. Therefore the original species are converted to more volatile, thermally stable derivatives. Alkylation is a common derivatisation method that can be utilized in speciation analysis. Sometimes hydride formation is also possible. When the derivatization reaction is carried out one has to aware that the original species remain unaltered, *i.e.* no substitution reactions occur.

A very common method to convert non-volatile species to volatile derivatives is ethylation with sodium tetraethylborate $NaB(C_2H_5)_4$. Sometimes also $NaBPh_4$ is used. Sodium tetraethylborate is a fairly stable reagent and the ethylation reaction can be carried out in aqueous solution. For example, to a few ml aqueous sample solution 1 ml of dichlorometane and 0.2 ml of acetate buffer (pH 5) is added. Then 0.2 ml of 1% (m/V) sodium tetraethylborate is introduced and the ethylation reaction is started. Since the ethylated species might be highly volatile the reaction cell must be carefully closed. After the reaction is complete, 1 µl of the separated organic phase is introduced into a capillary column of gas chromatograph and the chromatogram is recorder.

Another possibility to utilize alkylation reaction is the use of Grignard reagents (alkyl- or arylmagnesium halides). With these reagents methyl, ethyl, propyl, butyl, hexyl and phenyl derivatives can be prepared. The reaction must be carried out in non aqueous organic solvent. Therefore the ionic organometallic species must be first complexed (e.g. with dithizone or tropolone) and extracted to organic solvent before the reaction with Grignard reagent. When the reaction is complete the excess reagent must be destroyed with sulphuric acid prior to sample introduction to gas chromatograph.

In GC practically the whole gaseous sample enters the atomizer resulting in good sensitivity. However, only limited sample volumes can be introduced, especially when capillary columns are used with gas chromatography. If the detector coupled with GC is not sensitive enough (e.g. AAS), suitable preconcetration techniques are often needed, when for example environmental samples are analyzed. Several preconcentration methods are available, such as cryogenic trapping, purge and trap technique (PT) and solid phase microextraction. In purge and trap technique the volatile species are purged from the reaction mixture onto a solid adsorbent. After collection, the species are

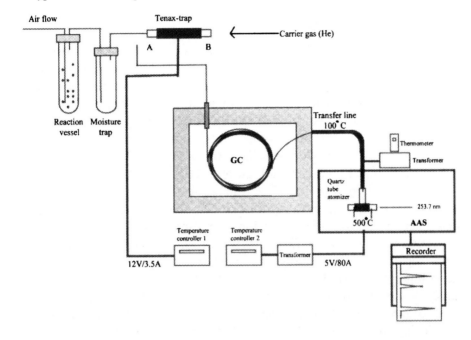

Figure 167 *Schematic of the PT-GC-QT-AAS equipment used for the determination of mercury species. The desorption temperature used for ethylated mercury species adsorbed on Tenax-ta adsorbent is 200°C*
(Adapted from: I. Välimäki and P. Perämäki, Microchim. Acta, 2001, **137**, 191)

thermally desorbed and introduced for gas chromatographic separation (Figure 167).

Hydride generation method has some applications as a derivatisation method in speciation studies. In addition to inorganic forms, some organic compounds of the hydride forming elements will form volatile hydrides. For example, monomethyl and dimethyl derivatives of arsonic acid will form the hydrides CH_3AsH_2 and $(CH_3)_2AsH$, respectively. Hydride generation reaction of different ionic methyl and butyl tin species has also some importance. Hydride derivatives are collected in a U-tube that is kept in liquid nitrogen (cryogenic trapping). The tube is filled with a suitable support material that is coated with an adsorbent (for example dimethylpolysiloxane). When the tube is heated, the adsorbed hydride derivatives are thermally desorbed according to their boiling points (thermal desorption gas chromatography). The resolution obtained with this inexpensive system is poor when compared to gas chromatography employing capillary columns, but is anyhow suitable for many environmental applications.

9.2.2 Liquid Chromatography

The disadvantage of GC separation is the need for volatile species/
derivatives. For instance, the analysis of large metal containing proteins
with GC is impossible. In many cases it is not possible to carry out the
derivatization reaction to obtain volatile compounds without altering
the chemical form of the original species (e.g. amino acids containing
selenium). The species or their derivatives may also be thermally
instable. In these cases the analysis with coupled techniques utilizing
high performance liquid chromatography (HPLC) in its various forms
is a possible alternative. In HPLC the separations are usually carried
out at room temperature. In contrast to GC, the constitution of mobile
phase has an important effect on the resolution obtained.

In speciation studies utilizing partition chromatography reversed-
phase separations are usually employed, *i.e.* the stationary phase of
the column is non-polar (often C_{18}) and the mobile phase is polar.
Aqueous mobile phase is buffered to suitable pH value and contains
polar organic solvents (e.g. acetonitrile and methanol). In addition,
complexing agents are often added to mobile phase to improve
separation. For instance when mercury species are determined, sulfur
containing ligands, such as 2-mercaptoethanol or cysteine, may be
added.

In ion-pair chromatography quite similar columns and conditions
are used than in reversed-phased partition chromatography. A large
organic anion (e.g. dodecylsulfonate) or a cation (e.g. tetrabutylammonium ion) is added (most often to the mobile-phase) to improve
the separation of the species. Large organic molecule interacts with
a stationary phase of the column and forms an ion pairs with the
oppositely charged analyte species. In the determination of arsenic
species tetrabutylammonium ion may be used as ion-pairing reagent
(Figure 168). The charges of the analyte species are sometimes modified
with additional complexing agents (e.g. F^-).

Ion-chromatography is also very often applied in speciation studies
employing HPLC technique. Both anion exchange and cation exchange
columns are used, depending on the charge of the analyte species. The
aqueous mobile mobile phase contains usually weak acid(s) and is
buffered to a suitable pH value. Water soluble organic solvents, such as
methanol, may be added to improve solubility of analytes. Separation is
based on a competition of the ion exchange sites of the resin between
the analyte species and eluent molecules. In the case of cations hydrated
proton competes from cation exchange sites and separations can be
controlled by adjustment of the pH value of the eluent. For instance,

alkylated cationic tin species can be separated by ion-exchange chromatography with a mobile phase containing ammonium citrate and methanol.

The most common HPLC separation methods mentioned above are convenient to use. However, the concentrations of the analyte species in many types of samples are very low and a limited sample volume (generally few tenths of microlitres) can be injected into the column. Since the mobile phase after HPLC separation is usually introduced to the detector by solution nebulisation having poor sample introduction efficiency, only ICP-MS is in practice a technique sensitive enough in many applications.

Sometimes the sample introduction efficiency may be improved by vapour generation (hydride generation, cold vapour method or sometimes ethylation). With vapour generation it is possible to use various atomic spectrometric detectors. Some arsenic species, for example, are capable of forming hydrides after HPLC-separation (Figure 168). In other cases, on line sample oxidation (microwave technique or UV photo oxidation) must be employed prior to hydride generation.

Spectral interferences can be avoided by chromatographic sample introduction. For example, when rare earths are determined by ICP-AES with direct solution nebulization, spectral interferences are often caused by other co-existing rare earth elements. When a HPLC is connected to an ICP-AES, these spectral interferences can be excluded

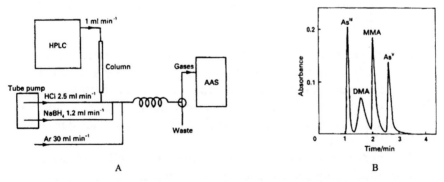

A   B

Figure 168 *Determination of arsenic species in urine samples. (A) HPLC-separation and on-line hydride generation prior arsenic atomization in a flame heated quartz tube. (B) Chromatogram of the 50 μl urine sample after separation with two ChromSep columns in series (ChromSpher C$_{18}$ packing). Mobile phase: 10 mmol l^{-1} tetrabutylammonium ion −20 mmol l^{-1} phosphate ion, pH 6.0 (range 6.0–5.4); MMA = monomethylarsonic acid, DMA = dimethylarsinic acid*
(Adapted from: E. Hakala and L. Pyy, J. Anal. Atom. Spectrom., 1992, 7, 191)

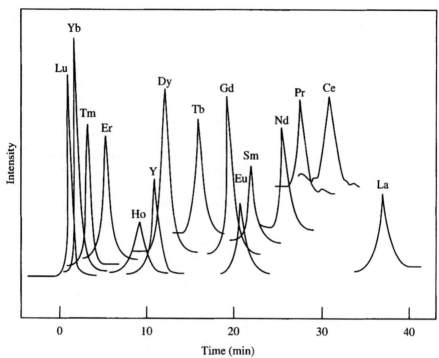

Figure 169 *Chromatograms for rare earth elements obtained by the HPLC-ICP-AES system. Conditions: 10 µg of each element in the sample; mobile phase (linear concentration gradient method), 0.4 M ammonium lactate, pH 4.22 (0–8 min), 0.6 M ammonium lactate, pH 4.22 (18 min), 1.0 M ammonium lactate, pH 4.22 (31–40 min)*
(Adapted from: K. Yoshida and H. Haraguchi, Anal. Chem., 1984, **56**, 2580)

by separating the rare earths by HPLC before their introduction into the plasma. Sodium or potassium salts of α-isobutyric acid, EDTA, or citric acid are generally used for the separation of rare earths are unsuitable elution reagents for the ICP-AES detection, as these compounds easily clog the nebulizer and reduce the nebulization efficiency because of their high viscosity.

Efficient separation of rare earths by the HPLC-ICP-AES system has been obtained using a linear concentration gradient elution method and ammonium lactate as the mobile phase (Figure 169). Detection limits for various rare earths vary between 0.001 (Yb) and 0.1 (Sm) mg l^{-1}, which are about an order of magnitude higher than those obtained by direct solution nebulization. However, the HPLC-ICP-AES method requires only 0.1 ml of the sample solution, while at least 1 ml of the sample solution is needed for the conventional ICP-AES method. In addition, the direct solution nebulization method suffers from spectral interferences, which are avoided by the HPLC separation in the HPLC-ICP-AES method. The calibration graphs showed linear relationships up to about 500 mg l^{-1} for all rare earth elements. RSD-values examined at 10 mg l^{-1} were about 4%. The method has been successfully applied in the determination of rare earths in different rock standard samples, rare earth ores, and high-purity lanthanide reagents.

CHAPTER 10

Advantages and Mutual Comparison of Atomic Spectrometric Methods

10.1 DETECTION LIMITS

Detection limits that can be achieved with various atomic spectrometric techniques are compiled in Table 49, and comparison of dynamic ranges of FAAS, GF-AAS, ICP-AES, and ICP-MS are shown in Figure 170. However, it must be remembered that method detection limits are usually considerably higher to the instrumental detection limits shown in Table 49, depending on the sample type analyzed and the possible interference effects existing with the particular technique used. Table 49 clearly shows that these techniques are complementary rather than competitive. GF-AAS is very important in the determination of metals at p.p.b.-level. Detection limits achieved by ICP-MS are even lower, most often at sub p.p.t. level. Elements which are difficult to detect by FAAS are generally easily determined by plasma AES. For example, detection limits achieved by FAAS for boron, lanthanum, niobium, phosphorus, tantalum, uranium, and tungsten are considerably higher than those achieved by ICP-AES or DCP-AES.

10.2 ATOMIC ABSORPTION

Atomic absorption methods possess a number of advantages: (i) high specificity; (ii) low detection limits; (iii) easy to use; (iv) low investment and running costs; (v) interferences and methods for their elimination are well documented; (vi) easy sample preparation; (vii) a number of special techniques available for the determination of non-metals and organic compounds. However, in practice AAS is still a single element method. One element is determined in a series of samples and the instrumental parameters are optimized for the next element and the series is repeated. Thus, simultaneous multi-element analysis is technically

Table 49 *Instrumental 3σ detection limits (μg l⁻¹) attainable for aqueous samples with various atomic spectrometric methods*
(Perkin Elmer Corp)

Element	FAAS	Hg/ Hydride	GF-AAS[b]	ICP-AES	ICP-MS
Ag	1.5		0.005	0.6	0.002
Al	45		0.1	1	0.005
As	150	0.03	0.05	2	0.0006[a]
Au	9		0.15	1	0.0009
B	1000		20	1	0.003
Ba	15		0.35	0.03	0.00002[a]
Be	1.5		0.008	0.09	0.003
Bi	30	0.03	0.05	1	0.0006
Br					0.2
C					0.8 (^{13}C)
Ca	1.5		0.01	0.05	0.0002[a]
Cd	0.8		0.002	0.1	0.00009[a]
Ce				1.5	0.0002
Cl					12
Co	9		0.15	0.2	0.0009
Cr	3		0.004	0.2	0.0002[a]
Cs	15				0.0003
Cu	1.5		0.014	0.4	0.0002
Dy	50			0.5	0.0001 (^{163}Dy)
Er	60			0.5	0.0001
Eu	30			0.2	0.00009
F					372
Fe	5		0.06	0.1	0.0003[a]
Ga	75			1.5	0.0002
Gd	1800			0.9	0.0008 (^{157}Gd)
Ge	300			1	0.001 (^{74}Ge)
Hf	300			0.5	0.0008
Hg	300	0.009	0.6	1	0.016 (^{202}Hg)
Ho	60			0.4	0.00006
I					0.002
In	30			1	0.0007
Ir	900		3.0	1	0.001
K	3		0.005	1	0.0002[a]
La	3000			0.4	0.0009
Li	0.8		0.06	0.3	0.001
Lu	1000			0.1	0.00005
Mg	0.15		0.004	0.04	0.0003
Mn	1.5		0.005	0.1	0.00007[a]
Mo	45		0.03	0.5	0.001
Na	0.3		0.005	0.5	0.0003
Nb	1500			1	0.0006
Nd	1500			2	0.0004
Ni	6		0.07	0.5	0.0004
Os				6	
P	75000		130	4	0.1
Pb	15		0.05	1	0.00004[a]
Pd	30		0.09	2	0.0005
Pr	7500			2	0.00009

Table 49   *Continued*

Element	FAAS	Hg/ Hydride	GF-AAS	ICP-AES	ICP-MS
Pt	60		2.0	1	0.002
Rb	3		0.03	5	0.0004
Re	750			0.5	0.0003
Rh	6			5	0.0002
Ru	100		1.0	1	0.0002
S				10	28 (^{34}S)
Sb	45	0.15	0.05	2	0.0009
Sc	30			0.1	0.004
Se	100	0.03	0.05	4	0.0007[a]
Si	90		1.0	10	0.03[a]
Sm	3000			2	0.0002
Sn	150		0.1	2	0.0005
Sr	3		0.025	0.05	0.00002[a]
Ta	1500			1	0.0005
Tb	900			2	0.00004
Te	30	0.03	0.1	2	0.0008 (^{125}Te)
Th				2	0.0004
Ti	75		0.35	0.4	0.003 (^{49}Ti)
Tl	15		0.1	2	0.0002
Tm	15			0.6	0.00006
U	15000			10	0.0001
V	60		0.1	0.5	0.0005
W	1500			1	0.005
Y	75			0.2	0.0002
Yb	8			0.1	0.0002 (^{173}Y)
Zn	1.5		0.02	0.2	0.0003[a]
Zr	450			0.5	0.0003

[a]Dynamic reaction cell (DRC) was employed.
[b]50 µl sample volume.

difficult to perform. In addition, qualitative analysis is impractical. A further disadvantage is that often dangerous gas mixtures must be used.

In principle, all elements may be determined on the basis of their ability to absorb light energy, but in practice this is not technically possible. There are several elements which have their strongest spectral lines in the vacuum UV region ($\lambda < 200$ nm), and thus makes their determination by direct AAS methods impossible (Table 50). On the other hand, some of these elements may be determined by using less sensitive absorption lines. For example, the sensitivity of the mercury resonance line at 186.96 nm is about 50 times more sensitive than the line at 253.65 nm used for the determination of mercury.

10.3   PLASMA ATOMIC EMISSION

Advantages of plasma atomic emission spectrometry are: (i) wide linear dynamic range (10^5–10^6 orders of magnitude); (ii) easy and rapid

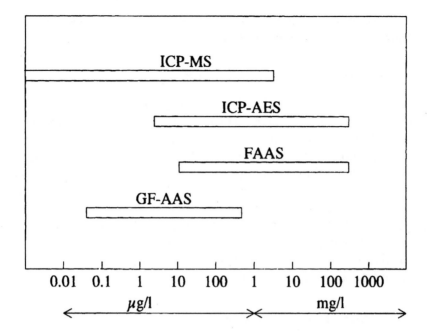

Figure 170 *Dynamic range of various atomic spectrometric methods*

Table 50 *Resonance lines located in the UV region*

Element	Resonance line (nm)	Element	Resonance line (nm)
F	95.2; 95.5	P	177.5; 178.3; 178.8
Cl	134.7; 138.0	S	180.7; 182.1; 182.6
Br	148.9; 157.6	As	189.0; 193.8; 193.7
I	183.0	Se	196.1
C	156.1;	Hg	184.9

qualitative analysis; (iii) simultaneous multi-element analysis; (iv) low running costs; (v) good precision, low detection limits, and high sensitivity (RSD values: FAAS 0.3–1%, GF-AAS 1–5%, ICP-AES 0.5–2%); (vi) minimized chemical interferences; (vii) analysis of more than 70 elements including refractories and some non-metals is possible (emission lines located in the vacuum UV region may be employed when measurements are performed in a vacuum or inert atmosphere); (viii) line rich spectra permit the use of several spectral lines for each element to be determined; (ix) plasma AES is safe because only inert gases are used. On the other hand, spectral interferences (direct coincidence of lines, overlapping of nearby lines, overlapping of broadened lines, and continuous radiation) are common in plasma AES.

10.4 PLASMA MASS SPECTROMETRY

Analytical benefits of ICP-MS are: (i) rapid multi-element analysis; (ii) rapid semiquantitative analysis which includes interpretation of spectra; (iii) very low detection limits; (iv) isotopic analysis including isotopic ratio (preferably with multicollector ICP-MS) and isotopic dilution analysis; (v) wide linear dynamic range ($>10^5$); (vi) spectral simplicity. ICP-MS shares analytical applications with plasma AES and AAS methods, multi-element capabilities with ICP-AES, and analytical speed with ICP-AES. On the other hand, ICP-MS is unique in isotopic measurement capabilities and in rapid semiquantitative analysis. The major disadvantage of quadrupole ICP-MS is the spectral interference caused by diatomic molecular ions.

10.5 ATOMIC FLUORESCENCE

In principle, AFS shares the advantages of atomic absorption and plasma atomic emission: (i) fewer spectral and chemical interferences; (ii) low detection limits; (iii) wide linear dynamic range. The main reason for its low popularity is the lack of commercial instruments. However, simple non-dispersive AF detectors are widely used in the determination of mercury and hydride forming elements.

CHAPTER 11

Further Reading

11.1 BOOKS

Several excellent general and topic-orientated textbooks have been written on atomic absorption, atomic emission spectroscopy and mass spectroscopy. Most of these are dealing with atomic absorption and/or inductively coupled plasma emission techniques:

G. F. Kirkbright and M. Sargant, 'Atomic Absorption and Fluorescence Spectroscopy', Academic Press, London, 1974.

W. J. Price, 'Spectrochemical Analysis by Atomic Absorption', Heyden, London, 1983.

M. Thompson and J. N. Walsh, 'Handbook of Inductively Coupled Plasma Spectrometry', Blackie, Glasgow, 1989.

B. Welz and M. Sperling, 'Atomic Absorption Spectrometry', Third, Completely Revised Edition, WILEY-VHC Verlag GmbH, Weinheim, 1999.

E. Metcalfe, 'Atomic Absorption and Emission Spectroscopy, Analytical Chemistry by Open Learning', ed. F. E. Prichard, John Wiley & Sons, Chichester, 1987.

'Inductively Coupled Plasmas in Analytical Atomic Spectrometry', Second Edition, ed. A. Montaser and D. W. Golightly, VCH Publishers Inc., New York, 1992.

'Inductively Coupled Plasma Mass Spectrometry', ed. A. Montaser, Wiley-VCH, New York, 1998.

'Plasma Source Mass Spectrometry: The New Millennium', ed. J. G. Holland and S. D. Tanner, The Royal Society of Chemistry, Cambridge, 2001.

'Plasma Source Mass Spectrometry: Applications and Emerging Technologies': ed. J. G. Holland and S. D. Tanner, The Royal Society of Chemistry, Cambridge, 2003.

Chemometrics is an important topic that has many applications in the field of analytical chemistry, for example in method development, calibration and evaluation of analytical data. The following textbook is a readable introduction to chemometrics:

J. N. Miller and J. C. Miller, 'Statistics and Chemometrics for Analytical Chemistry', Fourth Edition, Prentice Hall, Harlow, England, 2000.

11.2 GENERAL JOURNALS, SPECIALIST JOURNALS, ABSTRACTS, AND REVIEWS

11.2.1 General Journals

The following general analytical chemistry journals publish mainly original papers, but also sometimes reviews on various branches of analytical chemistry. Atomic spectrometric methods form an essential part of their contents:

Analytical Chemistry, American Chemical Society.
Analytica Chimica Acta, Elsevier.
The Analyst, The Royal Society of Chemistry.
Analytical and Bioanalytical Chemistry, Springer-Verlag.
Microchimica Acta, Springer-Verlag.
Zhurnal Analytiskoi Khimii.
Talanta, Elsevier

11.2.2 Specialist Journals

The following journals publish mainly original papers on special branches of analytical chemistry or only short communication:

Analytical Letters, Marcel Dekker.
Applied Spectroscopy, Society of Applied Spectroscopy.
Atomic Spectroscopy (earlier Atomic Absorption Newsletter), Perkin Elmer.
Canadian Journal of Analytical Sciences and Spectroscopy, Spectroscopy Society of Canada.
Journal of Analytical Atomic Absorption Spectrometry (JAAS), The Royal Society of Chemistry.
Spectrochimica Acta, Part B, Elsevier

11.2.3 Abstracts and Reviews

The following journals publish abstracts and reviews on analytical chemistry including atomic spectrometric methods:

Analytical Abstracts, The Royal Society of Chemistry.
Annual Reports on Analytical Atomic Spectroscopy, CRC Press
Applied Spectroscopy Reviews, Marcel Dekker.
Critical Reviews in Analytical Chemistry, Taylor & Francis
Progress in Analytical Atomic Spectroscopy, Pergamon Press.
Trends in Analytical Chemistry, Elsevier.

Subject Index

Lightning Source UK Ltd.
Milton Keynes UK
UKOW05n1507190514

231930UK00001B/51/A